Muscle Cell and Tissue - Novel Molecular Targets and Current Advances

Edited by Kunihiro Sakuma

Published in London, United Kingdom

IntechOpen

Supporting open minds since 2005

Muscle Cell and Tissue - Novel Molecular Targets and Current Advances
http://dx.doi.org/10.5772/intechopen.91539
Edited by Kunihiro Sakuma

Assistant to the Editor: Nobuo Morotomi

Contributors
Georgi P. P. Georgiev, Nobuo Morotomi, Kunihiro Sakuma, Kotomi Sakai, Yuichiro J. Suzuki, Nataliia V. Shults, Johnny Huard, Sudheer Ravuri, Naomasa Fukase, Ingrid K. Stake, Yoichi Murata, William S. Hambright, Marc J. Philippon, Yasmine F. Ibrahim, Vladyslava Rybka, Jaquantey R. Bowens, Adenike S. Falade, Aleš Fajmut

First published in London, United Kingdom, 2021 by IntechOpen
IntechOpen is the global imprint of INTECHOPEN LIMITED, registered in England and Wales, registration number: 11086078, 5 Princes Gate Court, London, SW7 2QJ, United Kingdom
Printed in Croatia

British Library Cataloguing-in-Publication Data
A catalogue record for this book is available from the British Library

Additional hard and PDF copies can be obtained from orders@intechopen.com

Muscle Cell and Tissue - Novel Molecular Targets and Current Advances
Edited by Kunihiro Sakuma
p. cm.
Print ISBN 978-1-83968-650-4
Online ISBN 978-1-83968-651-1
eBook (PDF) ISBN 978-1-83968-652-8

For EU product safety concerns:
IN TECH d.o.o., Prolaz Marije Krucifikse Kozulić 3, 51000 Rijeka, Croatia, info@intechopen.com or visit our website at intechopen.com.

Meet the editor

Professor Kunihiro Sakuma, Ph.D., currently works in the Institute for Liberal Arts at the Tokyo Institute of Technology. He is a physiologist working in the field of skeletal muscle. He was awarded his sports science diploma in 1995 by the University of Tsukuba and began his scientific work at the Department of Physiology, Aichi Human Service Center, focusing on the molecular mechanism of congenital muscular dystrophy and normal muscle regeneration. His interest later turned to the molecular mechanism and attenuating strategy of sarcopenia (age-related muscle atrophy). His opinion is to attenuate sarcopenia by improving autophagic defects using nutrient- and pharmaceutical-based treatments.

Contents

Preface

The contraction of skeletal muscles enables the body to move and maintain homeostasis. Human health is markedly affected by any deterioration in the material, metabolic, and contractile properties of skeletal muscle. The loss of skeletal muscle mass and strength substantially impairs physical performance and quality of life. Muscle wasting and weakness such as cachexia, atrophy, and sarcopenia are characterized by marked decreases in protein content, myonuclear number, muscle fiber size, and muscle strength. To attenuate various forms of muscle wasting, many researchers have investigated exercise-based, supplemental, and pharmacological strategies. This book introduces some approaches to the treatment of muscle wasting.

Our circulatory system is managed by the heart, lungs, and vasculature. These components serve crucial roles in controlling blood and lymph flow, and in the delivery of gases, hormones, and essential nutrients (i.e., glucose, fat, or amino acids). Pulmonary arterial hypertension (PAH) is a fatal disease without a cure. If untreated, increased pulmonary vascular resistance kills patients within several years due to right heart failure. This book also introduces novel applications against PAH such as cell reprogramming and the use of anticancer drugs that induce programmed cell death.

Vascular smooth muscle cells (VSMCs) are the most prevalent cell types in blood vessels and serve critical regulatory roles, particularly for vasoconstriction, vasodilatation, and synthesis of the vascular extracellular matrix. Cyclic guanosine 3′,5′-monophosphate (cGMP) is an intracellular second messenger that facilitates a broad spectrum of physiologic processes in multiple cell types within the cardiovascular, gastrointestinal, urinary, reproductive, nervous, endocrine, and immune systems. In particular, cGMP signaling plays a vital role in regulating the endothelium, VSMCs, and cardiac myocytes. This publication also reviews mathematical models concerning the molecular mechanism and the targets of cGMP in the contraction of VSMCs.

This book will be of interest to professionals in clinical practice, medical and health care students, and researchers working in muscle-related fields of science.

Kunihiro Sakuma
Professor,
Institute for Liberal Arts, Environment and Society,
Tokyo Institute of Technology,
Tokyo, Japan

Section 1

Therapeutic Approaches for Muscular Disorder

Chapter 1

Exercise Therapy for Patients with Heart Failure: Focusing on the Pathophysiology of Skeletal Muscle

Nobuo Morotomi, Kunihiro Sakuma and Kotomi Sakai

Abstract

In patients with heart failure (HF), it is important to perform exercise therapy with a focus on the pathophysiology of skeletal muscle. Patients with HF have multiple clinical symptoms due to cardiac dysfunction. Recent studies demonstrated the mechanism and treatment strategy for HF, and multiple signaling pathways involved in HF result in reduced exercise capacity and skeletal muscle mass. On the other hand, exercise therapy for HF is known to inhibit the inflammatory cytokines and neurohumoral factors, and increase muscle mass. Therefore, in this chapter, we discuss the importance of exercise therapy for HF, with a focus on the pathophysiology of skeletal muscle.

Keywords: heart failure, skeletal muscle, muscle abnormality, exercise training, combination therapy

1. Introduction

1.1 Pathology of HF

HF is a condition characterized by cardiac decompensation due to organic or functional impaired pumping capacity. Reduced exercise capacity in patients with HF leads to numerous symptoms such as breath shortness, dyspnea, general fatigue, and edema in the foot [1]. One of the causes of reduced exercise capacity is the condition of HF. According to the JCS 2017/JHFS 2017 guidelines on the diagnosis and treatment of acute and chronic HF [1], HF is categorized into three groups by the left ventricular ejection fraction (LVEF). One is HF with LVEF <40%, termed heart failure with reduced ejection fraction (HFrEF). HFrEF develops due to left ventricular systolic dysfunction. HFrEF is a leading cause of ischemic heart disease and coronary sclerosis. Another group is HF with LVEF ≥50%, termed heart failure with preserved ejection fraction (HFpEF). HFpEF is known as diastolic dysfunction in HF. Diastolic dysfunction is mainly caused by hypertension. The last group is HR with LVEF from 40–49%, termed heart failure with mid-range ejection fraction (HFmrEF). HFmrEF can develop in the recovery process from HFrEF, but many other factors can cause HFmrEF. For example, arrhythmia with tachycardia, such as atrial fibrillation, places strain on the left heart and causes HFmrEF. The exact mechanism of HFmrEF remains unclear.

Reduced exercise capacity in patients with HF can also cause organic and functional abnormality of skeletal muscle not only due to abnormal LVEF. Patients with HF can lose weight due to these abnormalities of skeletal muscle. A previous study reported that weight loss of 7.5% over six months was an independent predictor of long-term mortality in patients with HF [2]. In addition to the condition of HR and its related skeletal muscle abnormalities, reduced exercise capacity develops slowly with age. On the other hand, in 1998, Rosenberg proposed the term "sarcopenia" to describe age-related muscle decrease. Elderly patients with HF are included in "secondary sarcopenia" and advanced HF patients are defined as having "cardiac cachexia". The Asian definition of sarcopenia was established by the Asian Working Group for Sarcopenia in 2014 [3]. Sarcopenia is diagnosed by a low muscle mass and low muscle strength or low physical performance. Sarcopenia develops due to aging, undernutrition, sedentary lifestyle, and progression of inflammatory diseases, cancer, and chronic diseases such as HF and COPD. The reported prevalence of sarcopenia is between 4.1% and 11.5% in Asian older people. On the other hand, in 2013, Fülster reported a prevalence of 19.4% among 200 ambulatory patients with stable chronic HF in Germany [4].

The exact mechanism of the development of sarcopenia related to HF is unclear. However, previous studies noted specific changes in skeletal muscle in patients with HF. Kinugawa et al. reported that patients with HF have skeletal muscle abnormalities, including energy dysbolisms, fiber transformation from type 1 (*slow*-twitch muscle fibers) to type 2 (*fast*-twitch muscle fibers), and mitochondrial dysfunction [5]. Brown et al. demonstrated decreases in mitochondrial density, gene expression, and oxygenated capacity in skeletal muscle in patients with HF [6]. Using animal studies, Takada et al. found that chronic *heart failure* (CHF) model mice have a low level of protein expression of phosphorylated AMPKα, sirtuin-1, PGC-1, and mitochondrial transcription factor A (Tfam) in skeletal muscle [7].

In this review, we describe the molecular mechanisms in skeletal muscle and the treatment for HF.

2. Molecular mechanisms of HF

This section presents three representative skeletal muscle mechanisms unique to HF that cause organic and functional physical abnormalities.

2.1 Inflammatory cytokines

Inflammatory cytokines are biologically active substances that exert a variety of functions by signaling through specific receptors on the cell surface. Inflammatory cytokines include interleukin (IL)-6, interferon, tumor necrotic factor (TNF) family, and colony-stimulating factor. The excess secretion of inflammatory cytokines due to HF disrupts the anabolic catabolic balance of skeletal muscle in HF patients. Impaired cardiac cell and vascular endothelial cells with high stress result in the secretion of inflammatory cytokines and myocardial infarction. Levine et al. reported that HF patients have a high circulating concentration of TNF-α [8].

In 2001, Reid et al. reported that secretion of TNF-α increased the production of reactive oxygen species through the mitochondrial electron transport system, resulting in the activated ubiquitin-proteasome system through the NF-κB signaling pathway [9]. This study suggested progressive skeletal muscle wasting through these systems in HF patients. In addition, Chojkier et al. reported that

TNF-α-injected mice had increased expression of nitric oxide synthase activity and the expression of their albumin synthesis genes was downregulated [10]. Schaap et al. reported that *high* expression of TNF-α and IL-6 led to a reduction of the quadriceps muscle area and grip strength in older people [11]. Langhaus et al. reported that the secretion of TNF-α induced appetite loss [12]. Saitoh et al. found that appetite loss caused malnutrition and resulted in the inhibition of protein anabolic action. Moreover, appetite loss was associated with a poor prognosis in HF patients in their study [13]. On the other hand, Hollriegel et al. reported that HF patients did not have high expression of atrogin-1 mRNA or protein in skeletal muscle [14]. Of note, there are few studies on this topic in human subjects [15]. Although TNF-α and IL-6 activity affect muscle abnormalities in HF patients, it remains unclear how inflammatory cytokine families influence each other in humans.

2.2 Renin–angiotensin–aldosterone system

The accelerated renin-angiotensin-aldosterone (RAA) system leads to skeletal muscle dysfunction in HF patients. Angiotensinogen is produced from the liver and is the first substrate of the RAA system. Thereafter, renin from the juxtaglomerular apparatus of the kidney is released into the blood. Angiotensin I is produced from angiotensinogen by activated renin. Then, angiotensin II is produced from angiotensin I by angiotensin-converting enzyme. Angiotensin II causes vasoconstriction and stimulates aldosterone secretion through *angiotensin* II *type 1 receptor.* Angiotensin II causes an accumulation of fibrosis in skeletal muscle through transforming growth factor-β (TGF-β)-dependent signaling [16]. Activation of the RAA system in HF patients causes the overproduction of angiotensin II. Their RAA system is activated regardless of the severity of HF [17]. This study also revealed that their renin activity is significantly higher than that in those without HF (3.0 ± 3.7 and 1.2 ± 1.2 ng/mL). In 2007, Cohn et al. investigated the effects of TGF-β neutralizing antibodies and angiotensin-receptor blockers (ARB) on skeletal muscle function. In this study, the intervention group using ARB had significantly reduced TGF-β signaling through angiotensin II and improved skeletal muscle function [18]. TGF-β is also known as a factor that elicits apoptosis of muscle satellite cells during the process of repairing damaged skeletal muscles. The author suggested that inhibition of the angiotensin II type1 receptor by ARB led to the low expression of TGF-β.

Fukushima et al. demonstrated that the phosphorylation of Akt (p-Akt) decreased in skeletal muscle from mice with HF after myocardial infarction and this decrease was caused by an increase in plasma angiotensin II [19]. Another study using HF model mice found that angiotensin II directly induced skeletal muscle abnormalities [20]. This study reported a significant decrease in the amount of p-Akt protein in angiotensin II-treated mice. In addition, angiotensin II induced fiber transformation from type I to type II, skeletal muscle atrophy, and weight loss. Based on these involved factors, the RAA system is considered to elicit skeletal muscle dysfunction in HF. In the future, studies on increased protein catabolism by the RAA system need to be conducted in humans.

2.3 Autophagy

Autophagy is an intracellular system that delivers cytoplasmic substrates to lysosomes for subsequent degradation and removal. Autophagy is divided into three types depending on the mechanisms; macroautophagy, microautophagy, and chaperone-mediated autophagy. The research area of macroautophagy is the most advanced of these three types. Fasting induces the autophagy system and the isolation membrane is formed. The isolation membrane elongates, engulfing protein aggregates and organelles within the cytoplasm, and finally forms double-membraned structures called autophagosomes. Autophagosomes subsequently fuse with lysosomes to degrade their cargo by lysosomal enzymes.

Autophagy influences the muscle abnormalities in HF. There are two mechanisms. First, a maladaptive response for autophagy exacerbates the condition of HF, resulting in reduced muscle function. For example, the decrease in Beclin 1 expression weakens the macroautophagy system in patients with ischemic cardiac myopathy [21]. In contrast, Zhu et al. reported that pressure overload in mice markedly increased cardiac autophagy and load-induced autophagic activity remained significantly high for at least 3 weeks [22]. This study reported that Beclin 1 overexpression increased autophagic activity and promoted pathological remodeling. Using mice with pressure-overload heart failure, this study revealed that lysosome abundance calculated by measuring the lysosomal markers LAMP-1 and cathepsin D increased in wild-type hearts and to a greater extent in Beclin 1 transgenic hearts. Another study demonstrated that cardiac-specific deficiency of autophagy-related 5, a protein required for autophagy, leads to cardiac hypertrophy in adult mice [23]. Therefore, HF is exacerbated by dysfunctional autophagy and results in skeletal muscle abnormality.

Second, cardiac autophagy may directly cause skeletal muscle atrophy regardless of the progression of HF. Janning et al. investigated the autophagy pathway using myocardial infarction model mice [24]. Their study revealed that although myoatrophy in the soleus muscle and plantaris muscle progressed, the expression levels of autophagic markers, such as GABARAPL-1 and AtG7, increased in the plantaris but not in the soleus muscle. This study provides evidence of autophagy signaling regulation in HF-induced muscle atrophy. In addition, the selective degradation of mitochondria is termed mitophagy. Oka et al. reported that cardiac cells are abundant in mitochondria and dysfunctional mitophagy leads to reduced cardiac function through the inflammation inside the cells [25]. It is also possible that mitophagy causes skeletal abnormalities.

There is increasing evidence supporting a role of autophagy in age-related disease states of the cardiovascular system. Sasaki et al. reviewed autophagy in cardiovascular disease [26]. Their report states that autophagy is related to age-associated cardiovascular diseases, HF, ischemic heart disease, cardiomyopathy, hypertension, and atherosclerosis. However, the mechanisms of skeletal muscle dysfunction caused by autophagy in HF remain unclear.

3. Exercise training as treatment for HF

Exercise training improved the reduced exercise capacity and skeletal muscle power due to HF in several studies. There are two types of exercise; aerobic exercise and resistance training (RT). Aerobic exercise induced peroxisome proliferator-activated receptor gamma coactivator 1-alpha (PGC1-α) expression and improved insulin resistance [27]. High-intensity aerobic exercise increases the ratio of type 1 muscle fibers. Gielen et al. investigated how the expression of cathepsin-L, E3 ligases MuRF-1, and MaFbx changed after exercise training among HF patients, and compared them with healthy subjects [28]. As a result, the expression of MuRF-1 in HF patients was significantly higher than that in healthy subjects. In addition, after four weeks of exercise training, the expression of MuRF-1 mRNA in HF patients was reduced by 32.8% and 37.0% in people aged $\leq$55 years and $\geq$ 65 years, respectively. In another study, they investigated the expression of inflammatory cytokines (TNF-α, IL-6, and IL-1-β) before and after exercise training in HF patients [29].

Exercise training did not affect the serum levels of TNF-α, IL-6, or IL-1-β, but it significantly reduced the expression levels of these cytokines and iNOS (by 52%) in skeletal muscle. Thus, exercise training may reduce the expression of inflammatory cytokines and maintain high-level muscle function.

St-Jean-Pelletier et al. investigated myofiber changes and mitochondrial density in the vastus lateralis in healthy subjects [30]. They found an increased ratio of type 2a myofibers and decreased mitochondrial density in people aged ≥65 years with low physical activity. Campos et al. investigated the effects of exercise training on mitochondrial dysfunction using myocardial infarction model mice [31]. This study suggested that the improvement in mitochondrial number, density, and oxygenation by exercise training aid in recovery from cardiac dysfunction. Although exercise training improved mitochondrial dysfunction in HF mice model, the exact mechanism in HF patients remains to be elucidated.

Aerobic exercise increases the exercise capacity, and RT improves skeletal muscle mass and strength. Pu et al. demonstrated the effects of resistance training on muscle function in HF patients [32]. In their study, the improvement of knee extensor muscle power was 43% higher and the six-minute walking distance was 49 m greater in HF patients than those in the control group. Another study also found that RT improved skeletal muscle mass and power more than aerobic exercise in dialysis patients [33]. Multimodal exercise programs, including aerobic exercise, RT, and respiratory muscle training, were reported to significantly improve dyspnea and the quality of life, in addition to quadriceps power and exercise time, in HF patients [34].

Saitoh et al. suggested that combination therapy of exercise training with standard drugs, such as angiotensin-converting enzyme inhibitors, beta-receptor blockers, ghrelin agonists, and myostatin inhibitors, is better than exercise training and nutritional supplements for treating cardiac sarcopenia [35]. The following section explains the possibility of treatment for HF using combinations of exercise training and other therapies (**Figure 1**).

3.1 Combination of nutritional supplements and exercise training

3.1.1 Amino acids and micronutrients

Sufficient nutritional supplements improve skeletal muscle dysfunction accompanying HF. Aquilani et al. investigated the effects of exercise training with 8 g of daily essential amino acids (EAA) in 21 HF patients [36]. Based on the cardiopulmonary exercise test, the EAA group had no change in oxygen consumption but increased their exercise load. Although there was no significant change in the 6-minute walking distance in the control group, the EAA group increased their walked distance by 74 m on average. Rozentryt et al. reported that the intake of nutritional supplements with a high-calorie (600 kcal) and high-protein (20 g) diet increased the body weight by 2.0 kg at the 6-week follow-up and 2.3 kg at the 18-week follow-up [37]. In this study, oral nutritional supplementation did not affect the albumin concentration or peak oxygen consumption, but reduced the serum level of TNF-α.

β-Hydroxy-β-methylbutyrate (HMB) is a metabolite of the amino acid leucine and has a positive effect on muscle protein anabolism. A study using rats reported that HMB supplementation resulted in greater expression of Akt, mTOR, and S6K1 than leucine [38]. Berk et al. investigated the effects of a mixture of HMB, glutamine, and arginine in advanced cancer patients. However, there were no significant differences in the 8-week lean body mass between the placebo and the HMB/Arg/Gln groups [39]. Another study examining chronic pulmonary patients

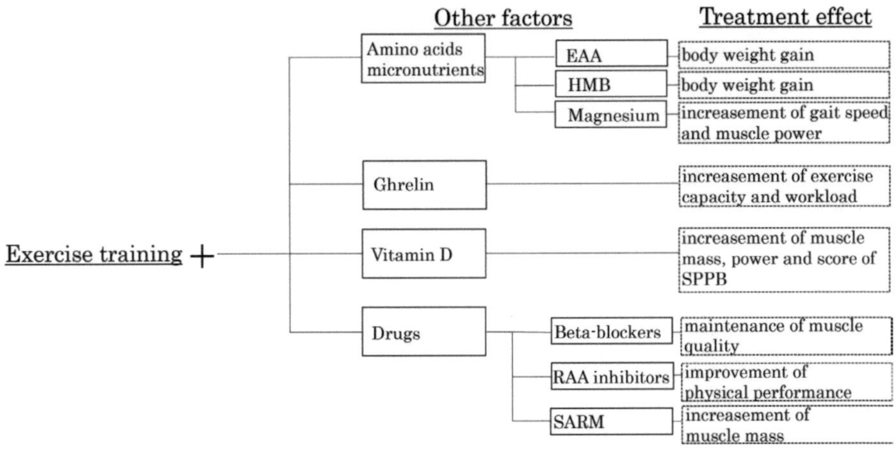

Figure 1.
It is thepossibility of treatment for HF using combinations of exercise training and other therapies. This figure presents the effects of exercise training added by other factors. The square box shows the other factors and the dotted box shows treatment effect. EAA: essential amino acids, HMB: β-Hydroxy-β-methylbutyrate, SPPB: short physical performance battery, RAA inhibitors: Renin-angiotensin-aldosterone inhibitors, SARM: selective androgen receptor modulators.

described that, in the group receiving pulmonary rehabilitation plus an oral nutritional supplement enriched with HMB, the mean and maximum handgrip and fat free mass significantly increased at 12 weeks [40]. The review of HMB supplementation in humans suggested that this agent has positive effects in patients with chronic pulmonary disease, hip fracture, and AIDS- related and cancer-related cachexia [41], but HF was not mentioned. However, a recent study suggested that a high-protein oral nutritional supplement containing HMB increased the body weight at day 30 in HF patients [42]. As described above, protein supplementation for HF patients increases the body mass and improves muscle function.

In recent years, the role of microelements has gained attention. Magnesium insufficiency increase the risk of HF. Sasiwarang et al. examined the relationship between the onset of HF and magnesium concentration in healthy subjects for 15 years [43]. Serum magnesium was inversely related to the risk of incident of HF. Moreover, HF patients with hypomagnesemia had high levels of IL-6 and von Willebrand factor (VWF). VWF is a marker for endothelial dysfunction and the serum level of VWF in HF patients is high. The author suggested that the magnesium concentration influences the inflammatory reaction. Microelements also play an important role in the treatment of sarcopenia [44]. Magnesium supplementation was reported to possibly improve the functional indices such as quadriceps torque [45]. In addition, the walking speed of healthy elderly women in the magnesium supplementation group became significantly faster than that of those in the control group (the supplementation group:Δ 0.21 ± 0.27 m/s, the control group Δ 0.14 ± 0.003). However, there are no studies on the effects of combination therapy of exercise training and magnesium supplementation in HF patients. As magnesium is commonly used for the treatment of arrhythmia and HF, it may be useful for the treatment of muscle dysfunction in HF.

3.1.2 Ghrelin

Ghrelin can improve the physical function of patients with HF. Ghrelin is produced in the fundic gland of the stomach, and stimulates *gastric* acid secretion and *motility.* Ghrelin has anabolic, orexigenic, and anti-inflammatory effects [46]. Ghrelin levels are lower in older people, especially in those with sarcopenia [47].

A study on anamorelin, a selective ghrelin receptor agonist, demonstrated a significant effect on body weight and food intake, but not on muscle strength in patients with cancer cachexia [48]. Nagaya et al. reported that the injection of synthesized ghrelin (2 μg/kg twice a day for 3 weeks) to HF patients increased the LVEF without adverse events, and increased the peak workload and oxygen consumption during exercise [49]. On the other hand, rikkunshito, a Japanese herbal medicine, is a ghrelin potentiator. Fujitsuka et al. reviewed the promotion of ghrelin activity by rikkunshito [50]. Several clinical trials demonstrated that the administration of rikkunshito increased the plasma ghrelin levels. These studies support the potential use of rikkunshito for improving skeletal muscle function and exercise capacity. However, rikkunshito and dipotassium glycyrrhizinate are structural components of licorice extract. The accumulation of dipotassium glycyrrhizinate in the body may cause pseudo aldosteronism and exacerbate the condition of HF. Thus, rikkunshito should be administered carefully. The satisfactory amount of rikkunshito should be investigated to manage HF effectively and safely.

3.1.3 Vitamin D

Vitamin D administration improves the exercise capacity in HF patients. The role of vitamin D is to maintain homeostatic function of the calcium-phosphorus balance and regulate bone metabolism. In recent years, vitamin D was confirmed to play an essential role in skeletal muscle function. Vitamin D deficiency or mutated vitamin D receptor causes skeletal muscle atrophy [51]. Vitamin D receptors are involved in gene expression in skeletal muscle, and regulate muscle anabolism and metabolism. The receptors act on calcium channels and directly regulate muscle contraction. Therefore, vitamin D deficiency results in lipid accumulation in skeletal muscle and atrophy of type 2 myofibers. Hayakawa et al. reported that the administration of 1α25 (OH)2D3 to human muscle cells inhibited the gene expression of MaFbx and MuRF-1 [52]. Antoniak et al. examined the effects of combination therapy of vitamin D administration and exercise training in comparison with exercise training and vitamin D alone [53]. They found that *lower extremity muscle power* increased more in the combination therapy group than in the exercise training alone group. In addition, the score of the short physical performance battery, skeletal muscle power, and femur density increased more in the combination therapy group than in the vitamin D alone group.

On the other hand, a meta-analysis demonstrated that vitamin D administration reduced the levels of TNF-α, CRP, and thyroid hormone, but did not improve exercise performance [54]. Bauer et al. reported the effects of combination therapy using vitamin D and leucine-enriched whey protein on physical function in older people with sarcopenia [55]. The active group (n = 184) received vitamin D at 800 IU, 20 g of whey protein, and 9 g of leucine twice a day for 13 weeks. In the active group, the score for the chair-stand test (1.0 second on average) and muscle mass (0.19 kg on average) significantly improved when compared with the control group. Other several studies using healthy elderly subjects reported the improvement of physical functions using a combination of vitamin D and amino acids, but not in HF patients.

3.2 Combination with standard therapeutics for HF

Standard therapeutics for HF can improve the skeletal function in HF patients. This section explains the three types of medicines for HF that may be useful for skeletal muscle.

3.2.1 Beta-blockers

Beta-blockers can prevent weight loss in HF patients. Beta-blockers, which have inhibitory action against left ventricular remodeling, are used for the treatment of HF and hypertension. Bisoprolol is a beta one-selective blocker and carvedilol is an alpha-beta blocker, and they are commonly used in the treatment of HF.

Beta-blockers were recently reported to inhibit muscle atrophy. A study using cancer cachexia model mice revealed that the administration of bisoprolol inhibited the loss of skeletal muscle mass [56]. This study also reported that bisoprolol improved physical activity and oral intake. In patients with rectal cancer or small cell carcinoma, the administration of espindolol twice a day improved their life prognosis and weight loss [57].

Clark et al. investigated the effects of the administration of carvedilol on body weight loss in patients with HF [58]. Carvedilol was initially administered at 3.1 mg (twice a day) and later increased to a maximum of 25 mg per dose (twice a day). As a result, the administration of carvedilol resulted in body weight gain (1 kg on average) after one year.

Based on these studies, beta-blockers may maintain skeletal muscle quality and improve skeletal muscle mass in HF patients.

3.2.2 Renin-angiotensin-aldosterone (RAA) inhibitors: angiotensin converting enzyme inhibitors (ACEI)/angiotensin-receptor blockers (ARB)

RAA inhibitors can improve muscle function in HF patients. Angiotensin converting enzyme inhibitors (ACEI) and angiotensin receptor blockers (ARB) are commonly used for the treatment of hypertension. Saitoh et al. reported the protective action of ACEI for muscle function [59]. Several studies reported that ACEI prevent body weight loss due to HF, and improved muscle power and physical functions [60, 61]. Sumukadas et al. investigated the effects of a combination of ACEI therapy and exercise training on physical functions in older people [62]. However, the 6-minute walk distance, score of the Short Physical Performance Battery, handgrip and quadriceps strength, and QOL at 10-week and 20-week follow-up did not improve in the intervention group. In this study, some bias was considered to have caused the insufficient effects of combination therapy.

RAA inhibitors have direct positive effects on muscle function in HF patients. Thus, high-quality studies examining both RAA inhibitors and exercise training in HF patients are warranted.

3.2.3 Selective androgen receptor modulators (SARM)

Testosterone may have a positive effect on skeletal muscle mass. On the other hand, serious side effects, such as increased risks of developing prostate cancer and myocardial infarction, have been reported [63]. Therefore, selective androgen receptor repair agents known as selective androgen receptor modulators (SARM) were developed. They exert testosterone-induced muscle mass gain effects with less stimulation of the prostate. SARM may increase skeletal muscle mass in HF patients. The clinical trial SARMsMK-0773 examined their effects on skeletal muscle in female patients with sarcopenia [64]. Muscle mass in the intervention group increased significantly compared with that in the placebo group, but no effects on physical functions or muscle power were observed in this study.

No studies have investigated the effects of combination therapy of SARM and exercise training on skeletal muscle mass and physical function in HF patients.

4. Conclusions

The pathophysiology of skeletal muscle in patients with HF is complex and remains unclear. However, recent studies clarified several points. The mechanisms do not function independently and instead interact with each other. In the management of HF, it is important to assess skeletal muscle and physical functions, and to consider treatment combinations including exercise training, electrical stimulation, medicine, and supplements. Large-scale, high-quality studies are warranted to elucidate the pathophysiology and establish effective treatments for HF patients.

Conflict of interest

I do not have any conflict of interest.

Author details

Nobuo Morotomi[1]*, Kunihiro Sakuma[2] and Kotomi Sakai[3]

1 Department of Rehabilitation medicine, Shin-Yurigaoka General Hospital, Kawasaki, Kanagawa, Japan

2 Institute for Liberal Arts, Environment and Society, Tokyo Institute of Technology, Tokyo, Japan

3 Setagaya Memorial Hospital, Department of Rehabilitation Medicine, Tokyo, Japan

*Address all correspondence to: uoratubon@gmail.com

References

[1] Tsutsui H, Isobe M, Ito H, Okumura K, Ono M, Kitakaze M, Kinugawa K, Kihara Y, Goto Y, Komuro I, Saiki Y, Saito Y, Sakata Y, Sato N, Sawa Y, Shiose A, Shimizu W, Shimokawa H, Seino Y, Node K, Higo T, Hirayama A, Makaya M, Masuyama T, Murohara T, Yano M, Yamazaki K, Yamamoto K, Yoshikawa T, Yoshimura M, Akiyama M, Anzai T, Ishihara S, Inomata T, Imamura T, Iwasaki Y, Ohtani T, Onishi K, Kasai T, Kobayashi S, Sakata Y, Tanaka A, Toda K, Noda T, Nochioka K, Hatano T, Fujino T, Makita S, Yamaguchi O, Ikeda U, Kimura T, Kohsaka S, Kosuge M, Yamagishi M, Yamashina A, Japanese Circulation society and the Japanese Heart Failure Society Joint Working Group (2019) JCS 2017/JHFS 2017 Guideline on Diagnosis and Treatment of Acute and Chronic Heart Failure-Digest Version. Circ J. 83: 2084-2184.

[2] Anker SD, Ponikowski P, Varney S, Chua TP, Clark AL, Webb-peploe KM, Harrigton D, Kox WJ, Poole-Wilson PA, Coats AJ (1997) Wasting as independent risk factor for mortality in chronic heart failure. Lancet. 349: 1050-1053.

[3] Chen LK, Liu LK, Woo J, Assantachai P, Auyeung TW, Bahyah KS, Chou MY, Chen LY, Hsu PS, Krairt O, Lee JSW, Lee Y, Liang CK, Limpawattana P, Lin CS, Peng LN, Satake S, Suzuki T, Won CW, Wu CH, Wu SN, Zeng P, Akishita M, Arai H (2014) Sarcopenia in Asia: consensus report of the Asian Working Group for Sarcopenia. J Am Med Disc Assoc. 15: 95-101.

[4] Fülster S, Tacke M, Sandek A, Ebner N, Tschope C, Doehner W, Anker SD, Haehling SV (2013) Muscle wasting in patients with chronic heart failure: results from the studies investigating co-morbidities aggravating heart failure (SICA-HF). Eur Heart J. 34: 512-519.

[5] Kinugawa S, Takada S, Matsushima S, Okita K, Tsutsui H (2015) Skeletal muscle abnormalities in Heart Failure. Int Heart J. 56: 475-484.

[6] Brown DA, Perry JB, Allen ME, Sabbah HN, Stauffer BL, Shaikh SR, Cleland JGF, Colucci WS, Butler J, Voors AA, Anker SD, Pitt B, Pieske B, Filippatos G, Greene SJ, Gheorgehiade M (2017) Expert consensus document: Mitochondrial function as a therapeutic target in heart failure. Nat Rev Cardiol. 14: 238-250.

[7] Takada S, Masaki Y, Kinugawa S, Matsumoto J, Furihata T, Mizushima W, Kadoguchi T, Fukushima A, Homma T, Takahashi M, Harashima S, Matsushita M, Yokota T, Tanaka S, Okita K, Tsutsui H (2016) Dipeptidyl peptidase-4 inhibitor improved exercise capacity and mitochondrial biogenesis in mice with heart failure via activation of glucagon-like peptide-1 receptor signalling. Cardiovasc Res. 111: 338-347.

[8] Levine B, Kalman J, Mayer L, Fillit HM, Packer M (1990) Elevated circulating levels of tumor necrosis factor in severe chronic heart failure. N Engl J Med. 323: 236-241.

[9] Reid MB, Li YP (2001) Tumor necrosis factor-alpha and muscle wasting: a cellular perspective. Respir Res. 2: 269-272.

[10] Chojkier M (2005) Inhibition of albumin synthesis in chronic diseases: molecular mechanisms. J Clin Gastroenterol. 39: S143-S146.

[11] Schaap LA, Pluijm SMF, Deeg DJH, Harris TB, Kritchevsky SB, Newman AB, Colbert LH, Pahor M, Rubin SM, Tylavsky FA, Visser M,

Health ABC study (2009) Higher inflammatory marker levels in older persons: associations with 5-year change in muscle mass and muscle strength. J Gerontol A Biol Sci Med Sci. 64: 1183-1189.

[12] Langhaus W, Hrupka B (1999) Interleukins and tumor factor as inhibitors of food intake. Neuropeptides. 33: 415-424.

[13] Saitoh M, Santos MPD, Ebner N, Emami A, Konishi M, Ishida J, Valentova M, Sandek A, Doehner W, Anker SD, Haehling SV(2016) Nutritional status and its effects on muscle wasting in patients with chronic heart failure: insights from Studies Investigating Co-morbidities Aggravating Heart Failure. Wien Klin Wochenschr. 128: 497-504.

[14] Höllriegel R, Beck EB, Linke A, Adams V, Möbius-Winkler S, Mangner N, Sandri M, Gielen S, Gutberlet M, Hambrecht R, Schuler G, Erbs S (2013) Anabolic effects of exercise training in patients with advanced chronic heart failure (NYHA III b): impact on ubiquitin-ligases expression and skeletal muscle size. Int J Cardiol. 167: 975-980.

[15] Sakuma K, Yamaguchi A. Molecular mechanisms controlling skeletal muscle mass. In: Sakuma K, editor. Muscle cell and Tissue. IntechOpen; 2015. p.143-170. DOI:10.5772/60876

[16] Kakutani N, Takada S, Nambu H, Matsumoto J, Furihata T, Yokota T, Fukushima A, Kinugawa S (2020) Angiotensin-converting-enzyme inhibitor prevents skeletal muscle fibrosis in myocardial infarction mice. Skelet Muscle. 10:11.

[17] Rouleau JL, Bichent D, Dagenais GR, Arnold JM, Parker JO, Bernstein V, Lamas G, Nadeau C (1993) Activation of neurohumoral systems in postinfarction left ventricular dysfunction. J Am Coll Cardiol. 22: 390-398.

[18] Cohn RD, Erp CV, Habashi JP, Soleimani AA, Klein EC, Lisi MT, Gamradt M, Rhys CM, Holm TM, Loeys BL, Ramirez F, Judge DP, Ward CW, Dietz HC (2007) Angiotensin II type 1 receptor blockade attenuates TGF-β induced failure of muscle regeneration in multiple myopathic states. Nat Med. 13: 204-210.

[19] Fukushima A, Kinugawa S, Takada S, Matsushima S, Sobrin MA, Ono T, Takahashi M, Suga T, Homma T, Masaki Y, Furihata T, Kadogushi T, Yokota T, Okita K, Tsutsui H (2014) (Pro) renin receptor in skeletal muscle is involved in the development of insulin resistance associated with postinfarct heart failure in mice. Am J Physiol Endocrinol Metab. 307: E503-E514.

[20] Kadoguchi T, Kinugawa S, Takada S, Fukushima A, Furihata T, Homma T, Masaki Y, Mizushima W, Nishikawa M, Takahashi M, Yokota T, Matsushima S, Okita K, Tsutsui H (2015) Angiotensin II can directly induce mitochondrial dysfunction, decrease oxidative fibre number and induce atrophy in mouse hindlimb skeletal muscle. Exp Physiol. 100: 312-322.

[21] Valentim L, Laurence KM, Townsend PA, Carroll CJ, Soond S, Scarabelli TM, Knight RA, Latchman DS, Stephanou A (2006) Urocortin inhibits Beclin1-mediated autophagic cell death in cardiac myocytes exposed to ischemia/reperfusion injury. J Mol Cell Cardiol. 40: 846-885.

[22] Zhu H, Tannous P, Johnstone JL, Kong Y, Shelton JM, Richardson JA, Le V, Levine B, Rothermel BA, Hill JA (2007) Cardiac autophagy is a maladaptive response to hemodynamic stress. J Clin Invest. 117: 1782-1793.

[23] Nakai A, Yamaguchi O, Takeda T, Higuchi Y, Hikoso S, Taniike M, Omiya S, Mizote I, Matsumura Y, Asahi M, Nishida K, hori M, Mizushima N, Otsu K (2007) The role of autophagy in cardiomyocytes in the basal state and in response to hemodynamic stress. Nat Med. 13: 619-624.

[24] Jannig PR, Moreira JBN, Bechara LRG, Bozi LHM, Bacurau AV, Moteiro AWA, Dourado PM, Wisloff U, Brum PC (2014) Autophagy signaling in skeletal muscle of infarcted rats. PLoS One. 9: e85820.

[25] Oka T, Hikoso S, Yamaguchi O, Taneike M, Takeda T, Tamai T, Oyabu J, Murakawa T, Nakayama H, Nishida K, Akira S, Yamamoto A, Komuro I, Otsu K (2012) Mitochondrial DNA that escapes from autophagy causes inflammation and heart failure. Nature. 485: 251-255.

[26] Sasaki Y, Ikeda Y, Iwabayashi M, Akasaki Y, Ohishi M (2017) The impact of autophagy on cardiovascular senescence and diseases. Int Heart J. 58: 666-673.

[27] Lira VA, Benton CR, Yan Z, Bonen A (2010) PGC-1alpha regulation by exercise training and its influences on muscle function and insulin sensitivity. Am J Physol Endocrinol Metab. 299: E145-E161.

[28] Gielen S, Sandri M, Kozarez I, Kratzsch J, Teupser D, Thiery J, Erbs S, Mangner N, Lenk K, Hambrecht R, Schuler G, Adams V (2012) Exercise training attenuates MuRF-1 expression in the skeletal muscle of patients with chronic heart failure independent of age: the randomized Leipzig Exercise Intervention in Chronic Heart Failure and Aging catabolism study. Circulation. 125: 2716-2727.

[29] Gielen S, Adams V, Mobius-Winkler S, Linke A, Erbs S, Yu J, Kempf W, Schubert A, Schuler G, Hambrecht R (2003) Anti-inflammatory effects of exercise training in the skeletal muscle of patients with chronic heart failure. J Am Coll Cardiol. 42: 861-868.

[30] St-Jean-Pelletier Félix, Pion CH, Leduc-Gaudet JP, Sgarioto N, Zovile I, Barbat-Artigas S, Reynaud O, Alkaterji F, Lemieux FC, Grenon A, Gaudreau P, Hepple RT, Chevalier S, Belanger M, Morais JA, Aubertin-Leheudre M, Gouspillou G (2017) The impact of aging, physical activity and pre-frailty on skeletal muscle phenotype, mitochondrial content, and intramyocellular lipids in men. J Cachexia Sarcopenia Muscle. 8: 213-228.

[31] Campos JC, Queliconi BB, Bozi LHM, Bechara LRG, Dourado PMM, Andres AM, Jannig PR, Gomes KMS, Zambelli VO, Rocha-Resende C, Gutaimosim S, Brum PC, Mochly-Rosen D, Gottlieb RA, Kowaltowski AJ, Ferrera JCB (2017) Exercise reestablishes autophagic flux and mitochondrial quality control in heart failure. Autophagy. 13: 1304-1317.

[32] Pu CT, Johnson MT, Forman DE, Hausdorff JM, Roubenoff R, Foldvari M, Fielding RA, Singh MA (2001) Randomized trial of progressive resistance training to counteract the myopathy of chronic heart failure. J Appl Physiol. 90: 2341-2350.

[33] Rhee CM, Kalantar ZK (2014) Resistance exercise: an effective strategy to reverse muscle wasting in hemodialysis patients ? J Cachexia Sarcopenia Muscle. 5: 177-180.

[34] Lauotaris ID, Adamopoulos S, Manginas A, Panagiotakos DB, Kallistratos MS, Doulaptsis C, Kouloubinis A, Voudris V, Pavlides G, Cokkinos DV, Dritsas A (2013) Benefits of combined aerobic/resistance/ inspiratory training in patients with

chronic heart failure. A complete exercise model? A prospective randomized study. Int J Cardiol. 167: 1967-1972.

[35] Saito M, Ebner N, Haehling SV, Anker SD, Springer J (2018) Therapeutic considerations of sarcopenia in heart failure patients. Expert Rev Cardiovasc Ther. 16: 133-142.

[36] Aquilani R, Opasich C, Gualco A, Verri M, Testa A, Pasni E, Viglio S, Iadarola P, Pastoris O, Dossena M, Boschi F (2008) Adequate energy-protein intake is not enough to improve nutritional and metabolic status in muscle depleted patients with chronic heart failure. Eur Heart Fail. 10: 1127-1135.

[37] Rozentryt P, Haehling SV, Lainscak M, Nowak JU, Kalantar ZK, Polonski L, Anker SD (2010) The effects of a high-caloric protein rich oral nutritional supplement in patients with chronic heart failure and cachexia on quality of life, body composition, and inflammation markers: a randomized, double-blind pilot study. J Cachexia Sarcopenia Muscle. 1: 35-42.

[38] Giron MD, Vilchez JD, Salto R, Manzano M, Sevillano N, Campos N, Argiles JM, Rueda R & Lopez-Pedrosa JM (2016) Conversion of leucine to β-hydroxy-β-methylbutyrate by α-keto isocapriate dioxygenase is required for a potent stimulation of protein synthesis in L6 rat myotubes. J Cachexia Sarcopenia Muscle. 7: 68-78.

[39] Berk L, James J, Schwartz A, Hug E, Mahadevan A, Samuels M, Kachnic L (2008) A randomized, double-blind, placebo-controlled trial of a β-hydroxy-β-methylbutyrate, glutamine, and arginine mixture for the treatment of cancer cachexia (RTOG 0122). Support Care Cancer. 16: 1179-1188.

[40] Olveria G, Olveria C, Dona E, Palenque FJ, Porras N, Dorado A, Godoy AM, Rubio-Martinez E, Rojo-Martinez G, Martin-Valeno R (2016) Oral supplement enriched in HMB combined with pulmonary rehabilitation improves body composition and health related quality of life in patients with bronchiectasis (Prospective, Randomized study). Clin Nutr. 35: 1015-1022.

[41] Holeček M (2017) Beta-hydroxy-beta-methylbutyrate supplementation and skeletal muscle in healthy and muscle wasting conditions. J Cachexia Sarcopenia Muscle. 8: 529-541.

[42] Deutz NE, Matheson EM, Matarese LE, Luo M, Baggs GE, Nelson JL, Hegazi RA, Tappenden KA, Ziegler TR, on behalf of the NOURISH Study Group (2016) Readmisson and mortality in malnourished, older, hospitalized adults treated with a specialized oral nutritional supplement: A randomized clinical trial. Clinical Nutrition. 35: 18-26.

[43] Wannamethee SG, Papacosta O, Lennon L, Whincup PH (2018) Serum magnesium and risk of incident heart failure in older men: The British Regional Heart Study. Eur J Epidemiol. 33: 873-882.

[44] Dronkelaar CV, Velzen AV, Abdelrazed M, Steen A VD, Weijs P JM, Tieland M (2018) Minerals and Sarcopenia; The Role of Calcium, Iron, Magnesium, Phosphorus, Potassium, Selenium, Sodium, and Zinc on Muscle mass, Muscle strength and Physical performance in older adults: A systematic review. J Am Med Dir Assoc. 19: 6-11.

[45] Zhang Y, Xun P, Wang R, Mao L, He K (2017) Can magnesium enhance exercise performance? Nutrients. 9: 946.

[46] Müller TD, Nogueiras R, Andermann ML, Andrews ZB, Anker SD, Argente J, Batterham RL, Benoit SC, Bowers CY, Broglio F,

Casanueva FF, D' Alessio D, Deportere I, Geliebter A, Ghigo E, Cole PA, Cowley M, Cummings DE, Dagher A, Siano S, Dickson SL, Dieguez C, Granata R, Grill HJ, Grove K, Habegger KM, Heppner K, Heiman ML, Holsen L, Holst B, Inui A, Janson JO, Kirchner H, Korbonits M, Laferrere B, LeRoux CW, Lopez M, Morin S, Nakazato M, Nass R, Perez-Tilve D, Pfluger PT, Schwartz TW, Seeley RJ, Sleeman M, Sun Y, Sussel L, Tong J, Thorner MO, van der Lely AJ, van der Ploeg LHT, Zigman JM, Kojima M, Kangawa K, Smith RG, Horvath T, Tschop MH (2015) Ghrelin. Mol Metab. 4: 437-460.

[47] M Serra-Prat, M Papiol, E palomera (2015) Relationship between plasma ghrelin levels and sarcopenia in elderly subjects: A cross-sectional study. J Nutr Health Aging. 19: 669-672.

[48] Temel J (2016) Anamorelin in patients with non-small-cell lung cancer and cachexia (ROMANA1 and ROMANA2): results from two randomized, double blind, phase 3 trials. Lancet Oncol. 17: 519-531.

[49] Nagaya N, Moriya J, Yasumura Y, Uematsu M, Ono F, Shimizu W, Ueno K, Kitakaze M, Miyatake K, Kanagawa K (2004) Effects of ghrelin administration on left ventricular function, exercise capacity, and muscle wasting in patients with chronic heart failure. Circulation. 110: 3674-3679.

[50] Fujitsuka N, Uezono Y (2014) Rikkunshito, a ghrelin potentiator, ameliorates anorexia-cachexia syndrome. Front Pharmacol. 5: 271.

[51] Molina P, Carrero JJ, Bover J, Chauveau P, Mazzaferro S, Torres PU (2017) Vitamin D, a modulator of musculoskeletal health in chronic kidney disease. J Cachexia Sarcopenia Muscle. 8: 686-701.

[52] Hayakawa N, Fukumura J, Yasuno H, Fujimoto-Ouchi K, Kitamura H (2015) 1α,25(OH)2D3 downregulates gene expression levels of muscle ubiquitin ligase MAFbx and MuRF1 in human myotubes. Biomed Res. 36: 71-80.

[53] Anatoniak AE, Greig CA (2017) The effect of combined resistance exercise training and vitamin D3 supplementation on musculoskeletal health and function in older adults: a systematic review and meta-analysis. BMJ Open. 7: e014619.

[54] Jiang WL, Gu HB, Zhang YF, Xia QQ, Qi J, Chen JC (2016) Vitamin D supplementation in the treatment of chronic heart failure: A Meta-analysis of randomized controlled trials. Clin Cardiol. 39: 56-61.

[55] Bauer JM, Verlaan S, Bautmans I, Brandt K, Donini LM, Maggio M, McMurdo M, Mets T, Seal C, Wijers SL, Ceda GP, Vito GD, Donders G, Drey M, Greig C, Holmback U, Narici M, McPhee J, Poggiogalle E, Power D, Scafoglieri A, Schultz R, Sieber CC, Cederholm T (2015) Effects of a vitamin D and leucine-enriched whey protein nutritional supplement on measures of sarcopenia in older adults, the PROVIDE study: a randomized, double-blind, placebo-controlled traial. J Am Med Dir Assoc. 16: 740-747.

[56] Springer J, Tscirner A, Haghikia A, Haehling S von, Lal H, Grzesiak G, Pauls S, Potsch M, Websky K von, H Berthold, Latouche C, Jaisser F, Morawietz L, Coats A J.S, Beadle J, Argiles JM, Thum T, Foldes G, Doehner W, Hilfiker-Kleiner D, Force T, Anker SD (2014) Prevention of liver cancer cachexia-induced cardiac wasting and heart failure. Eur Heart J. 35: 932-941.

[57] Coats AJ, Fuang HG, Prabhash K, Haehling S von, Tilson J, Brown R, Beadle J, Anker SD & for and on behalf of the ACT-ONE study group (2016) Espindlol for the treatment and

prevention of cachexia in patients with stage III/IV non-small cell lung cancer or colorectal cancer: a randomized, double-blind, placebo-controlled, international multicntre phase II study (the ACT-ONE trial). J Cachexia Sarcopenia Muscle. 7: 355-365.

[58] Clark AL, Krum H, Katus HA, Mohacsi P, Salekin D, Schultz MK, Packer M, Anker SD (2017) Effect of beta-adrenergic blockade with carvedilol on cachexia in severe chronic heart failure: results from the COPERNICUS trial. J Cachexia Sarcopenia Muscle. 8: 549-556.

[59] Saitoh M, Ebner N, Haehling SV, Anker SD, Springer J (2018) Therapeutic considerations of sarcopenia in heart failure patients. Expert Rev Cardiovasc Ther. 16: 133-142.

[60] Anker SD, Negassa A, Coats AJ, Poole-Wilson PA, Cohn JN, Yusuf S (2003) Prognostic importance of weight loss in chronic heart failure and the effect of treatment with angiotensin-converting-enzyme inhibitors: an observational study. Lancet. 361: 1077-1083.

[61] Zhou LS, Xu LJ, Wang XQ, Huhang YH, Xiao Q (2015) Effect of angiotensin-converting-enzyme inhibitors on physical function in elderly subjects: A systematic review and meta-analysis. Drugs Aging. 32: 727-735.

[62] Sumukadas D, Band M, Miller S, Cvoro V, Witham M, Struthers A, McConnachie A, Lloyd SM, McMurdo M (2014) Do ACE inhibitors improve the response to exercise training in functionally impaired older adults? A randomized controlled trial. J Gerontol A Biol Sci Med Sci. 69: 736-743.

[63] Borst SE, Shuster JJ, Zou B, Ye F, Jia H, Wokhlu A, Yarrow JF (2014) Cardiovascular risks and elevation of serum DHT vary by route of testosterone administration: a systematic review and meta-analysis. BMC Med. 12: 211.

[64] Papanicolaou DA, Ather SN, Zhu H, Zhou Y, Lutkiewicz J, Scott BB, Chandler J (2013) A phase IIA randomized, placebo-controlled clinical trial to study the efficacy and safety of the selective androgen receptor modulator (SARM), MK-0773 in female participants with sarcopenia. J Nutr Health Aging. 17: 533-543.

Chapter 2

Evidence for the Role of Cell Reprogramming in Naturally Occurring Cardiac Repair

Nataliia V. Shults and Yuichiro J. Suzuki

Abstract

Pulmonary arterial hypertension (PAH) is a fatal disease without a cure. If untreated, increased pulmonary vascular resistance kills patients within several years due to right heart failure. Even with the currently available therapies, survival durations remain short. By the time patients are diagnosed with this disease, the damage to the right ventricle (RV) has already developed. Therefore, agents that repair the damaged RV have therapeutic potential. We previously reported that cardiac fibrosis that occurs in the RV of adult Sprague–Dawley rats with PAH could naturally be reversed. We herein investigated the mechanism of this remarkable cardiac repair process. Counting of cardiomyocytes showed that the elimination of cardiac fibrosis is associated with the increased RV myocyte number, suggesting that new cardiomyocytes were generated. Immunohistochemistry showed the expression of α-smooth muscle actin and Sox-2 in RV myocytes of rats with PAH. Transmission electron microscopy detected the structure that resembles maturing cardiomyocytes in both the RV of PAH rats and cultured cardiomyocytes derived from induced pluripotent stem cells. We propose that the damaged RV in PAH can be repaired by activating the cell reprogramming mechanism that converts resident cardiac fibroblasts into induced cardiomyocytes.

Keywords: cardiac repair, cardiomyocyte regeneration, cell reprogramming, pulmonary hypertension, right ventricle

1. Introduction

Pulmonary arterial hypertension (PAH) affects males and females of any age, including children. Despite the availability of approved drugs, PAH remains a fatal disease without a cure [1, 2]. The major pathogenic features that increase the pulmonary vascular resistance in PAH include the vasoconstriction and the development of vascular remodeling, in which pulmonary artery (PA) walls are thickened and the lumens are narrowed or occluded. Increased resistance puts strain on the right ventricle (RV), and right heart failure is the major cause of death among PAH patients [3, 4]. The median overall survival for patients diagnosed with PAH is 2.8 years from the time of diagnosis (3-year survival: 48%) if untreated [5, 6]. Even with currently available therapies, the prognosis remains poor with only 58–75% of PAH patients surviving for 3 years [7–10]. PAH is a progressive disease; and by the time patients are diagnosed, RV damage has often already occurred.

RV failure is the major cause of death among patients with PAH. However, no treatment strategies are available to manage the RV dysfunctions. Physiologically, the RV needs to cope with fold changes in PA pressure. Thus, the RV is capable of adapting to increased pressure. Similarly to human patients with PAH, we found that the RV of Sprague–Dawley (SD) rats treated with SU5416 and hypoxia to produce PAH suffer from severe cardiac fibrosis at 8 to 17 weeks after the initiation of the SU5416/hypoxia treatment [11]. Remarkably, at 35 weeks after the initiation of the SU5416/hypoxia treatment, RV fibrosis was found to be resolved in these rats, despite the RV pressure remained high [12]. Thus, the RV remodeling can naturally be reversed in these animals, providing an interesting model of RV repair. Understanding the mechanism of such naturally occurring events should shed a light on developing therapeutic strategies to repair the cardiac damage in human patients. The present study examined the mechanism of this cardiac repair process.

2. Materials and methods

2.1 Experimental animals

Male adult SD rats (Charles River Laboratories International, Inc., Wilmington, MA, USA) were subcutaneously injected with SU5416 (20 mg/kg body weight; MedChem Express, Monmouth Junction, NJ, USA), maintained in hypoxia for 3 weeks [11, 12] and then in normoxia for up to 32 weeks (35-week time points). Animals were subjected to hypoxia in a chamber (30"w x 20"d x 20"h) regulated by an OxyCycler Oxygen Profile Controller (Model A84XOV; BioSpherix, Redfield, NY, USA) set to maintain 10% O_2 with an influx of N_2 gas, located in the animal care facility at the Georgetown University Medical Center [11, 12]. Ventilation to the outside of the chamber was adjusted to remove CO_2, such that its level did not exceed 5000 ppm. Control animals were subjected to ambient 21% O_2 (normoxia) in another chamber. Animals were fed normal rat chow. Animals were anesthetized and euthanized by excising the heart and the lungs.

The Georgetown University Animal Care and Use Committee approved all animal experiments, and the investigation conformed to the National Institutes of Health (NIH) Guide for the Care and Use of Laboratory Animals.

2.2 Immunohistochemistry (IHC)

RV tissues were immersed in buffered 10% formalin at room temperature, and were embedded in paraffin. These paraffin-embedded tissues were cut and mounted on glass slides. IHC was performed using horseradish peroxidase (HRP) labeled polymer and 3,3′-diaminobenzidine (DAB) chromagen (Agilent Technologies, Santa Clara, CA, USA) with α-smooth muscle actin (αSMA; Catalog # ab32575) and Sox2 (Catalog # ab97959) antibodies (Abcam, Cambridge, UK).

2.3 Transmission electron microscopy (TEM)

The RV free wall tissues of rats subjected to SU5416/hypoxia as well as cultured cardiomyocytes derived from human induced pluripotent stem cells (iPSCs) purchased from Cell Applications, Inc. (San Diego, CA) were fixed in the 2.5% glutaraldehyde/0.05 M cacodylate solution, post-fixed with 1% osmium tetroxide and embedded in EmBed812. Ultrathin sections (70 nm) were post-stained with uranyl acetate and lead citrate and examined in the Talos F200X FEG transmission electron microscope (FEI, Hillsboro, OR, USA) at 80 KV located at the George Washington University Nanofabrication and Imaging Center. Digital electron micrographs were recorded with the TIA software (FEI).

2.4 Statistical analysis

Means and standard errors were calculated. Comparisons between three groups were analyzed by using one-way analysis of variance (ANOVA) with a Student–Newman–Keuls post-hoc test using the GraphPad Prism (GraphPad Software, Inc., La Jolla, CA, USA). $P < 0.05$ was considered to be significant.

3. Results

Since the initial discovery that treating SD rats with the SU5416 injection plus chronic hypoxia promoted severe PAH with pulmonary vascular lesions that resemble those of human patients [13], this experimental model has become a gold standard in the research field of PAH. Experimental design usually involves the single subcutaneous injection of SU5416, followed by subjecting to chronic hypoxia for 3 weeks. In many studies, rats are then maintained in normoxia for 2 to 5 weeks, and PAH and pulmonary vascular remodeling are observed [13–17]. At this stage, some laboratories including ours have reported that the RV is severely damaged with fibrosis [11, 14, 18, 19]. At 8 to 17 weeks after the SU5416 injection, however, we found that, despite the occurrence of severe fibrosis, the RV contractility is maintained or even improved, perhaps due to the formation of 'super' RV myocytes [11]. Moreover, our more recent results demonstrated that, at 35 weeks (3 weeks hypoxia followed by 32 weeks of normoxia), RV fibrosis was largely resolved, despite the RV pressure remained high [12]. The number of myofibroblasts that contribute to the formation of fibrosis, as detected using a well utilized marker αSMA, is increased in non-vessel regions of the RV of SD rats with severe PAH as well as RV fibrosis, and this expression declined in rats with repaired RV at 35 week after the SU5416 injection [12]. Thus, the SU5416/hypoxia treatment does not cause the death of SD rats and the damaged RV can be repaired naturally. Since RV myocytes were not hypertrophied at 35-weeks, the number of myocytes must have been increased to fill the post-fibrotic areas. Indeed counting cardiomyocytes indicated that, compared to the 17-week time point when fibrosis was present, the number of cardiomyocytes increased at 35-weeks when the RV was repaired (**Figure 1**).

Based on these results, we hypothesize that these SD rats possess a mechanism that repairs the damaged heart in response to PAH. This raises a question on where regenerated cardiomyocytes come from when the RV repair occurs. It is now known that some adult cardiomyocytes are capable of proliferating [20]. However, the cardiac renewal through the cardiomyocyte proliferation would be too slow to replace large fibrotic areas that were seen in our experimental model. Another possibility is that new cardiomyocytes were regenerated from cardiac progenitor cells. We have identified c-kit and isl1-positive cardiac progenitor cells in the RV of SD rats, however, their levels were not altered in PAH rats (data not shown).

Our previous experiments described in Zungu-Edmondson et al. [12] using the αSMA antibody were originally performed for the purpose of detecting myofibroblasts, which express αSMA. By examining αSMA IHC slides from RVs of SD rats with PAH (at 17-weeks) with a larger magnification (x1,000) as shown in **Figure 2A** of Zungu-Edmondson et al. [12], we noticed that some brown

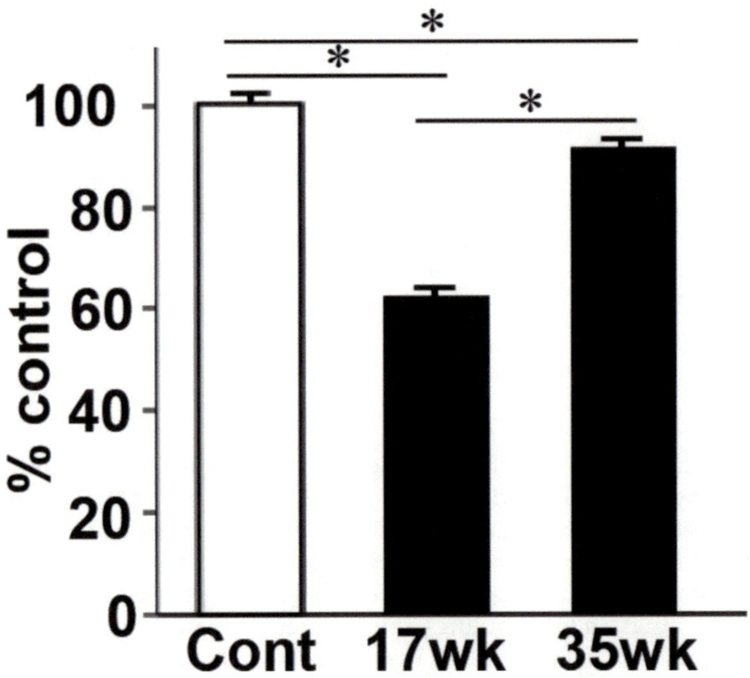

Figure 1.
*Restoration of RV cardiomyocytes. SU5416-injected SD rats were subjected to 3 weeks hypoxia and then maintained in normoxia to promote PAH. 17 and 35 weeks after the SU5416 injection, RV myocardium tissues were fixed in formalin, embedded in paraffin, and subjected to H&E staining. The number of cardiomyocytes were counted and expressed as % of the control rats (Cont). SD rats with PAH at 17 weeks after the initiation of SU5416/hypoxia had significantly reduced RV myocyte number compared to healthy controls. This decrease in RV myocyte number was restored at 35 weeks. The symbol * denotes that the values are significantly different from each other at p < 0.05.*

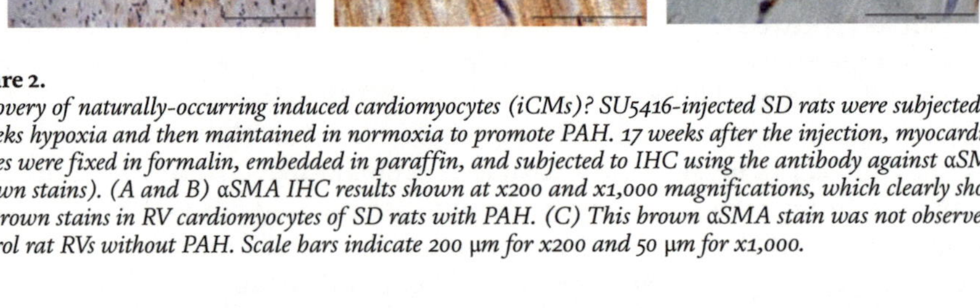

Figure 2.
Discovery of naturally-occurring induced cardiomyocytes (iCMs)? SU5416-injected SD rats were subjected to 3 weeks hypoxia and then maintained in normoxia to promote PAH. 17 weeks after the injection, myocardium tissues were fixed in formalin, embedded in paraffin, and subjected to IHC using the antibody against αSMA (brown stains). (A and B) αSMA IHC results shown at x200 and x1,000 magnifications, which clearly show the brown stains in RV cardiomyocytes of SD rats with PAH. (C) This brown αSMA stain was not observed in control rat RVs without PAH. Scale bars indicate 200 μm for x200 and 50 μm for x1,000.

αSMA stains, in addition to myofibroblasts indicated by arrows, also occurred on cardiomyocytes. We initially discarded these observations by thinking that they are non-specific artifacts. However, further examinations of a number of IHC slides made us convinced that cardiomyocytes are indeed stained with the αSMA antibody. This seems to occur regionally as a group in the myocardial walls (**Figure 2A**). **Figure 2B** shows the amplified view of **Figure 2A** clearly demonstrating that these cardiomyocytes with clear striations express αSMA. By contrast, control RVs from healthy rats without PAH did not exhibit the staining of cardiomyocytes with the αSMA antibody (**Figure 2C**).

Cell reprogramming is defined as the conversion of one specific cell type to another. This technology has gained considerable attention when Prof. Shinya Yamanaka discovered the means to convert fibroblasts into iPSCs and received the Nobel Prize [21]. In their study, stem cell-related transcription factors including Oct4 and Sox2 were used to convert somatic cells to pluripotent cells that can be differentiated into various cell types including cardiomyocytes [22]. More recently, a combination of cardiac-specific transcription factors was found to directly convert fibroblasts into cardiomyocytes [23, 24]. **Figure 3A** shows our TEM study of cardiomyocytes derived from iPSCs. These induced cardiomyocytes (iCMs) are capable of beating, express contractile proteins, and exhibit an organized sarcomere structure with clear striations and Z-lines (**Figure 3A**). In addition, some regions with the not well-defined sarcomere organization, which could be in the process of maturing into iCMs were also identified in these TEM images (**Figure 3B**). iCMs have been shown to express αSMA [25], while normal adult cardiomyocytes do not. Thus, αSMA-positive cardiomyocytes we observed in adult SD rats with PAH could be iCM-like cells. Consistently with this hypothesis, our examination of TEM images revealed that the structure resembling cardiomyocytes maturing from iPSCs as observed in cultured cells (**Figure 3B**) also occurs in the RV of SD rats with PAH (**Figure 4B**).

These results would support the concept that RVs of SD rats with PAH have iCMs-like cells that are produced via cell reprogramming. However, the αSMA expression can be induced by the fetal gene program mechanism that is associated with cardiac hypertrophy, independent of cell reprogramming. Thus, we examined if factors more directly related to cell reprogramming are expressed in these cardiomyocytes. We found that cardiomyocytes of RV tissues from SD rats with PAH also express Sox2 (**Figure 5A** and **B**), a stem cell-related transcription factor that has been used to generate iPSCs [22]. By contrast, no Sox2 stains were observed in healthy control SD rats without PAH (**Figure 5C**).

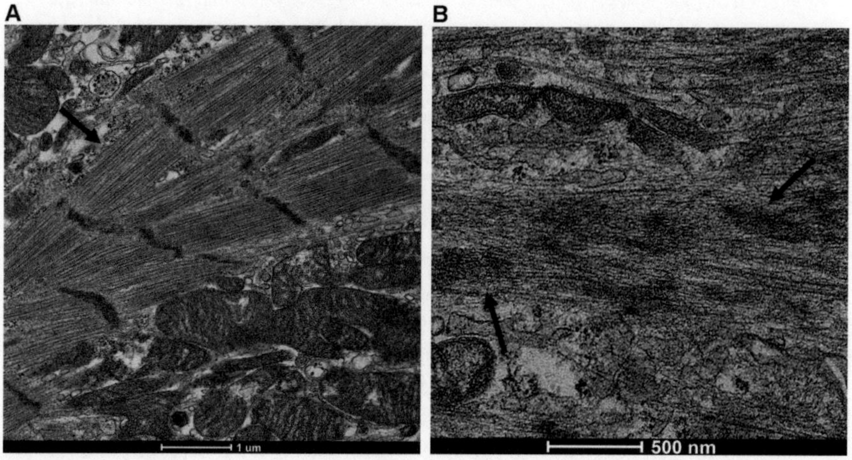

Figure 3.
TEM image of cardiomyocytes derived from iPSCs. The reprogramming technology was used to convert cultured human fibroblasts into iPSCs, then to cardiomyocytes (cell applications, Inc.). Fixed iPSC-derived cardiomyocytes were observed under a transmission electron microscope. (A) the representative image shows cardiomyocytes with clear striations and sarcomere structures (arrow), indicating that iPSCs indeed can become matured cardiomyocytes. Magnification, x5,500. (B) we also identified some regions with the not well defined sarcomere organization (arrows), which may be in the process of maturing into cardiomyocytes. Magnification, x14,000.

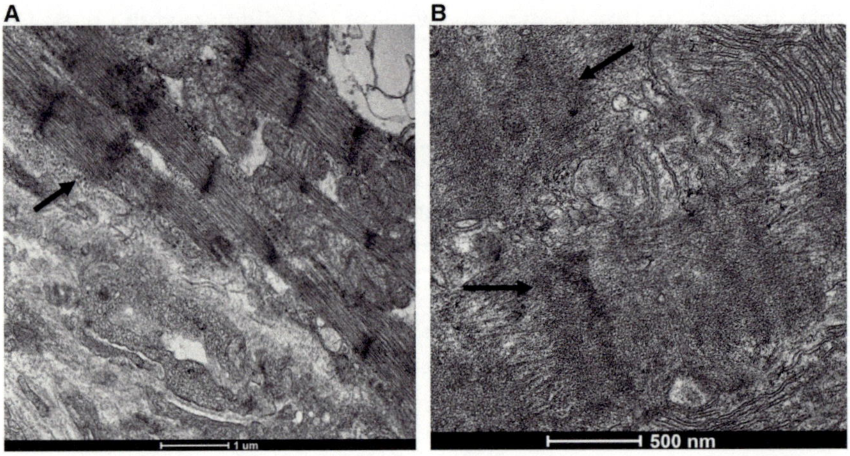

Figure 4.
TEM identification of maturing ICM-like cells in the RV of PAH rats. SD rats were injected with SU5416, subjected to 3 weeks hypoxia, and maintained in normoxia. 20 weeks after the injection, RV tissues were fixed and analyzed by TEM. (A) the image shows normal cardiomyocytes with clear striations and sarcomere structures (arrow). Magnification, x5,500. (B) we also identified some regions with the not well defined sarcomere organization (arrows), which may be in the process of maturing into iCMs, similar to the structure shown in ***Figure 3B****. Magnification, x14,000.*

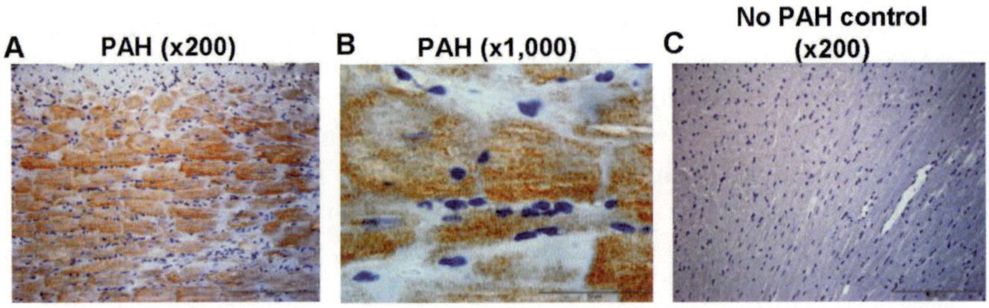

Figure 5.
Expression of Sox2. SU5416-injected SD rats were subjected to 3 weeks hypoxia and then maintained in normoxia to promote PAH. 17 weeks after the injection, hearts were fixed and subjected to IHC with the antibody against Sox2 (brown stains). (A and B) Sox2 IHC results shown at x200 and at x1,000 magnifications, which clearly show the brown stains in RV cardiomyocytes of SD rats with PAH. (C) This brown Sox2 stain was not observed in control rat RVs without PAH. Scale bars indicate 200 μm for x200 and 50 μm for x1,000.

4. Discussion

In response to pressure overload, the heart ventricles undergo a series of adaptive events. In response to systemic and pulmonary hypertension, the left ventricle (LV) and the RV, respectively hypertrophy in order to increase the force of muscle contraction. Concentric hypertrophy is the first change that occurs in response to chronic pressure overload, and this compensatory mechanism allows for improved cardiac output. Exercise-induced cardiac hypertrophy, for example, increases the force of contraction in accordance the needs associated with strenuous exercise and training. This adaptive feature is reversible. However, in chronic disease conditions, this compensatory mechanism thickens the ventricular wall too much in a manner that decreases the stroke volume and thus the cardiac output. This results in the second adaptation to decrease the ventricular wall thickness. In case of the LV, the transition from concentric to eccentric hypertrophy predominates, resulting in the dilated LV with the thin ventricular wall. This event, however, does not seem to occur in the RV in response to chronic pulmonary hypertension and the hallmark of cor pulmonale is that the RV myocytes remain concentrically hypertrophied at the time of heart failure. In the RV, the attempt to decrease the RV wall thickness is expected to be related to the promotion of cell death, although it is unclear whether apoptosis really occurs in the RV of human patients with pulmonary hypertension [26]. However, apoptotic cardiomyocytes have been detected in the RV of a rat model of pulmonary hypertension [11]. In these rats, regions of the RV where cardiomyocytes die get filled with fibrosis.

Using well-studied model of PAH, in which SD rats are treated with the SU5416 injection and chronic hypoxia, we previously found that RVs were capable of maintaining sufficient force of muscle contraction even severe fibrosis occurred at 8 to 17 weeks after the initiation of the SU5416/hypoxia treatment [11]. We postulated that this is due to the formation of "super RV myocytes" that are capable of eliciting stronger force of contraction via a mechanism involving the downregulation of calsequestrin 2, the major Ca^{2+}-binding protein of the sarcoplasmic reticulum.

Further, remarkably at 35 weeks after the SU5416/hypoxia initiation, these fibrosis regions disappear and are filled with newly formed cardiomyocytes [12]. The present study indeed showed that the number of cardiomyocytes were increased at the 35-week time point, compared to the 17-week time point. This increased number of cardiomyocytes was found to be associated with the production of RV cardiomyocytes that express αSMA and Sox2 as well as the occurrence of the structure visible in TEM that resembles maturing iCMs similar to those observed in cultured cardiomyocytes derived from iPSCs. From these results, we hypothesize that the damaged RV due to PAH can naturally be repaired in SD rats through a mechanism that involves the conversion of resident cardiac fibroblasts into iCMs via cell reprogramming (**Figure 6**). Our finding also suggests that the nature possesses the ability for maturing iCMs into functional cardiomyocytes that are capable of eliciting strong muscle contraction through a "functionalization" process (**Figure 6**). Our results so far provided evidence to support this novel mechanism of cardiac regeneration, however, further work is needed to prove this concept.

Understanding of the endogenous means to repair the heart is important, not only to provide basic knowledge of cardiac physiology, but also to develop therapeutic strategies to treat conditions in which cardiomyocytes are damaged. Along with the cardiomyocyte proliferation and the involvement of cardiac progenitor cells, our results suggest a novel mechanism of cardiac regeneration through cell reprogramming. Defining this naturally occurring mechanism should contribute to the development of technologies to convert resident cardiac fibroblasts into cardiomyocytes to save lives.

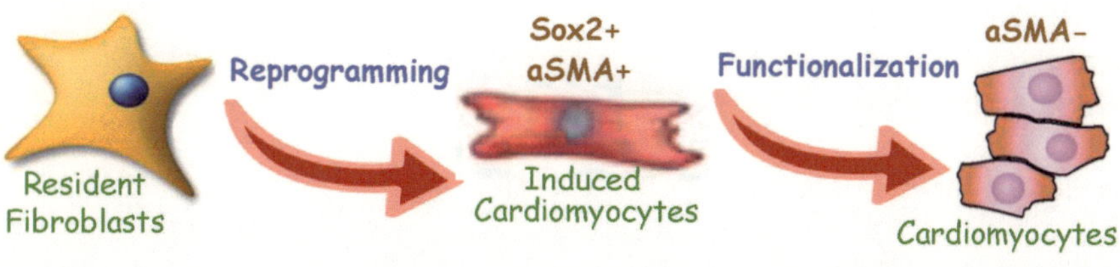

Figure 6.
Scheme depicting our hypothesis. SD rats possess an RV repair mechanism, in which naturally occurring cell reprogramming process converts resident fibroblasts into αSMA-and Sox2-positive induced cardiomyocytes (iCMs). These rats also possess the functionalization mechanism that makes functional cardiomyocytes that can elicit strong muscle contraction.

5. Conclusion

The present study generated evidence to support a fascinating and novel mechanism of naturally occurring cardiac repair that should be important for the development of effective therapeutic strategies to treat cardiac failure. This novel mechanism involves the conversion of resident cardiac fibroblasts into iCMs through the cell reprogramming process that appears to share events occurring in iPSC biology.

Acknowledgements

This work was supported in part by the NIH (grant numbers R01HL072844, R21AI142649, R03AG059554, and R03AA026516) to Y.J.S. The content is solely the responsibility of the authors and does not necessarily represent the official views of the NIH.

Author details

Nataliia V. Shults and Yuichiro J. Suzuki*
Department of Pharmacology and Physiology, Georgetown University Medical Center, Washington DC, USA

*Address all correspondence to: ys82@georgetown.edu

References

[1] Fallah F. Recent strategies in treatment of pulmonary arterial hypertension, a review. Glob J Health Sci. 2015;7:307-322.

[2] Rosenkranz S. Pulmonary hypertension 2015: current definitions, terminology, and novel treatment options. Clin Res Cardiol. 2015;104: 197-207.

[3] Delcroix M, Naeije R. Optimising the management of pulmonary arterial hypertension patients: emergency treatments. Eur Respir Rev. 2010;19:204-211.

[4] McLaughlin VV, Shah SJ, Souza R, Humbert M. Management of pulmonary arterial hypertension. J Am Coll Cardiol. 2015;65:1976-97.

[5] D'Alonzo GE, Barst RJ, Ayres SM, Bergofsky EH, Brundage BH, Detre KM, Fishman AP, Goldring RM, Groves BM, Kernis JT, et al. Survival in patients with primary pulmonary hypertension. Results from a national prospective registry. Ann Intern Med. 1991;115:343-349.

[6] Runo JR, Loyd JE. Primary pulmonary hypertension. Lancet. 2003;361:1533-1544.

[7] Benza RL, Miller DP, Frost A, Barst RJ, Krichman AM, and McGoon MD. Analysis of the lung allocation score estimation of risk of death in patients with pulmonary arterial hypertension using data from the REVEAL Registry. Transplantation. 2010;90:298-305.

[8] Humbert M, Sitbon O, Yaïci A, Montani D, O'Callaghan DS, Jaïs X, Parent F, Savale L, Natali D, Günther S, Chaouat A, Chabot F, Cordier JF, Habib G, Gressin V, Jing ZC, Souza R, Simonneau G; French Pulmonary Arterial Hypertension Network. Survival in incident and prevalent cohorts of patients with pulmonary arterial hypertension. Eur Respir J. 2010;36:549-555.

[9] Thenappan T, Shah SJ, Rich S, Tian L, Archer SL, Gomberg-Maitland M. Survival in pulmonary arterial hypertension: a reappraisal of the NIH risk stratification equation. Eur Respir J. 2010;35:1079-1087.

[10] Olsson KM, Delcroix M, Ghofrani HA, Tiede H, Huscher D, Speich R, Grünig E, Staehler G, Rosenkranz S, Halank M, Held M, Lange TJ, Behr J, Klose H, Claussen M, Ewert R, Opitz CF, Vizza CD, Scelsi L, Vonk-Noordegraaf A, Kaemmerer H, Gibbs JS, Coghlan G, Pepke-Zaba J, Schulz U, Gorenflo M, Pittrow D, Hoeper MM. Anticoagulation and survival in pulmonary arterial hypertension: results from the Comparative, Prospective Registry of Newly Initiated Therapies for Pulmonary Hypertension (COMPERA). Circulation. 2014;129:57-65.

[11] Zungu-Edmondson M, Shults NV, Wong CM, Suzuki YJ. Modulators of right ventricular apoptosis and contractility in a rat model of pulmonary hypertension. Cardiovasc Res. 2016;110:30-39.

[12] Zungu-Edmondson M, Shults NV, Melnyk O, Suzuki YJ. Natural reversal of pulmonary vascular remodeling and right ventricular remodeling in SU5416/hypoxia-treated Sprague-Dawley rats. PLoS One. 2017;12:e0182551.

[13] Taraseviciene-Stewart L, Kasahara Y, Alger L, Hirth P, Mc Mahon G, Waltenberger J, Voelkel NF, Tuder RM. Inhibition of the VEGF receptor 2 combined with chronic hypoxia causes cell death-dependent pulmonary endothelial cell proliferation and severe pulmonary hypertension. FASEB J. 2001;15:427-438.

[14] Oka M, Homma N, Taraseviciene-Stewart L, Morris KG, Kraskauskas D, Burns N, Voelkel NF, McMurtry IF. Rho kinase-mediated vasoconstriction is important in severe occlusive pulmonary arterial hypertension in rats. Circ Res. 2007;100:923-929.

[15] Ibrahim YF, Wong CM, Pavlickova L, Liu L, Trasar L, Bansal G, Suzuki YJ. Mechanism of the susceptibility of remodeled pulmonary vessels to drug-induced cell killing. J Am Heart Assoc. 2014;3:e000520.

[16] Alzoubi A, Toba M, Abe K, O'Neill KD, Rocic P, Fagan KA McMurtry IF, Oka M. Dehydroepiandrosterone restores right ventricular structure and function in rats with severe pulmonary arterial hypertension. Am J Physiol Heart Circ Physiol. 2013;304:H1708-H1718.

[17] Wang X, Ibrahim YF, Das D, Zungu-Edmondson M, Shults NV, Suzuki YJ. Carfilzomib reverses pulmonary arterial hypertension. Cardiovasc Res. 2016;110:188-199.

[18] Bogaard HJ, Natarajan R, Henderson SC, Long CS, Kraskauskas D, Smithson L, Ockaili R, McCord JM, Voelkel NF. Chronic pulmonary artery pressure elevation is insufficient to explain right heart failure. Circulation. 2009;120:1951-1960.

[19] Bogaard HJ, Natarajan R, Mizuno S, Abbate A, Chang PJ, Chau VQ, Hoke NN, Kraskauskas D, Kasper M, Salloum FN, Voelkel NF. Adrenergic receptor blockade reverses right heart remodeling and dysfunction in pulmonary hypertensive rats. Am J Respir Crit Care Med. 2010;182:652-660.

[20] Yuan X, Braun T. Multimodal regulation of cardiac myocyte proliferation. Circ Res. 2017;121:293-309.

[21] Yamanaka S. Induced pluripotent stem cells: Past, present, and future. Cell Stem Cell. 2012;10:678-684.

[22] Takahashi K, Yamanaka S. Induction of pluripotent stem cells from mouse embryonic and adult fibroblast cultures by defined factors. Cell. 2006;126:663-676.

[23] Ieda M, Fu JD, Delgado-Olguin P, Vedantham V, Hayashi Y, Bruneau BG, Srivastava D. Direct reprogramming of fibroblasts into functional cardiomyocytes by defined factors. Cell. 2010;142:375-386.

[24] Song K, Nam YJ, Luo X, Qi X, Tan W, Huang GN, Acharya A, Smith CL, Tallquist MD, Neilson EG, Hill JA, Bassel-Duby R, Olson EN. Heart repair by reprogramming non-myocytes with cardiac transcription factors. Nature. 2012;485:599-604.

[25] Cao N, Huang Y, Zheng J, Spencer CI, Zhang Y, Fu JD, Nie B, Xie M, Zhang M, Wang H, Ma T, Xu T, Shi G, Srivastava D, Ding S. Conversion of human fibroblasts into functional cardiomyocytes by small molecules. Science. 2016;352:1216-1220.

[26] Voelkel NF, Quaife RA, Leinwand LA, Barst RJ, McGoon MD, Meldrum DR, Dupuis J, Long CS, Rubin LJ, Smart FW, Suzuki YJ, Gladwin M, Denholm EM, Gail DB; Right ventricular function and failure: report of a National Heart, Lung, and Blood Institute working group on cellular and molecular mechanisms of right heart failure. Circulation. 2006;114:1883-1891.

Chapter 3

Interventional Strategies to Delay Aging-Related Dysfunctions of the Musculoskeletal System

Naomasa Fukase, Ingrid K. Stake, Yoichi Murata, William S. Hambright, Sudheer Ravuri, Marc J. Philippon and Johnny Huard

Abstract

Aging affects bones, cartilage, muscles, and other connective tissue in the musculoskeletal system, leading to numerous age-related pathologies including osteoporosis, osteoarthritis, and sarcopenia. Understanding healthy aging may therefore open new therapeutic targets, thereby leading to the development of novel approaches to prevent several age-related orthopaedic diseases. It is well recognized that aging-related stem cell depletion and dysfunction leads to reduced regenerative capacity in various musculoskeletal tissues. However, more recent evidence suggests that dysregulated autophagy and cellular senescence might be fundamental mechanisms associated with aging-related musculoskeletal decline. The mammalian/mechanical target of Rapamycin (mTOR) is known to be an essential negative regulator of autophagy, and its inhibition has been demonstrated to promote longevity in numerous species. Besides, several reports demonstrate that selective elimination of senescent cells and their cognate Senescence-Associated Secretory Phenotype (SASP) can mitigate musculoskeletal tissue decline. Therefore, senolytic drugs/agents that can specifically target senescent cells, may offer a novel therapeutic strategy to treat a litany of age-related orthopaedic conditions. This chapter focuses on osteoarthritis and osteoporosis, very common debilitating orthopaedic conditions, and reviews current concepts highlighting new therapeutic strategies, including the mTOR inhibitors, senolytic agents, and mesenchymal stem cell (MSC)-based therapies.

Keywords: Stem cells, senescence, mTOR, Osteoarthritis, Osteoporosis, aging

1. Introduction

Aging is a process of progressive loss of physiological function and reserve, characterized by cellular senescence, stem cell exhaustion, DNA damage, telomere attrition, and deregulated nutrient sensing [1]. It is well known that aging-related stem cell depletion and dysfunction leads to reduced tissue regenerative capacity in various stem cell populations. Osteoarthritis (OA) and osteoporosis (OP) are two common aging-related skeletal diseases that influence the quality of life in the elderly.

Osteoarthritis is one of the most common chronic degenerative joint diseases that cause joint pain and dysfunction. Multiple factors, including mechanical, genetic, and aging-related factors, are involved in the development of OA [2]. At the cellular level, it is characterized by loss of tissue cellularity, and phenotypic changes to chondrocytes, and subsequent damage to the extracellular matrix (ECM) [3]. Chondrocytes are resident cells in cartilage tissue and are involved in both the synthesis and turnover of the ECM [4]; therefore, maintaining chondrocyte health is an essential factor in preventing articular cartilage degeneration. With age, articular cartilage ECM degrades and remodels with the fragmentation of the principal proteoglycan protein aggrecan, chondroitin sulfate, and relative increases in keratan sulfate and hyaluronan deposition [5, 6]. Chondrocytes also exhibit an autolytic phenotype causing the proteolytic cleavage of various collagen molecules [7]. Collectively, such ECM perturbations during aging lead to a concomitant decrease in water content, affecting tensile strength of the cartilage and transmission of load.

Osteoporosis is a systemic bone degenerative disease characterized by a progressive loss of bone mass accompanied by a significant reduction in the mechanical strength of the bones [8]. Bone loss during aging is also characterized by changes in bone shape, mineral content, adiposity, mineral turnover, and reduction in bone forming osteoblasts [9, 10]. Another bi-product of aging is a reduction of bone healing capacity exacerbated by chronic inflammation inherent with advancing age [11] that also disrupts homeostatic interactions between signaling factors and bone cells, resulting in a state of dysregulated remodeling and bone turnover. Thus, aging is one of the factors most closely associated with the development of OP [12]. Brittle bones, as a result of OP in the elderly have been known to cause fractures triggered by minor trauma, significantly worsening the quality of life and even reduce life expectancy [13].

Recent research is beginning to unravel the mechanisms of how aging makes bones and joints more susceptible to the development of these diseases. Understanding the underlying mechanisms may provide new treatments to delay or prevent the development of these aging-related orthopedic disorders. This chapter focuses on OA and OP among aging-related dysfunctions of the skeletal system and reviews current concepts on new therapeutic strategies, especially the mammalian/mechanistic target of Rapamycin (mTOR) inhibitors, senolytic agents, and mesenchymal stem cells (MSCs).

2. Autophagy as a therapeutic target in aging-related orthopedic diseases

2.1 Autophagy and mTOR in OA

Autophagy is an essential homeostatic process for the clearance of damaged intracellular components [14]. It is conserved evolutionarily across species and is commonly known to be activated under stress conditions, including nutrient-deficiency, to generate energy [15]. When cells face nutritional stress, the autophagy pathway is activated to degrade unnecessary intracellular components while maintaining only the minimum essential components to prevent energy loss. Recent evidence has linked autophagy in the pathophysiology of OA. Inhibition of autophagy has been reported to promote the expression of OA-like genes induced by interleukin-1β (IL-1β), whereas activation of autophagy reduces the intracellular reactive oxygen species (ROS) by removing damaged mitochondria, thereby protecting chondrocytes from OA-like changes [16]. During aging and especially in OA, autophagic flux is diminished, likely due to decreases in autophagy regulator genes ULK1, beclin-1, and LC [17], and is associated with increased chondrocyte apoptosis and mitochondrial dysfunction [18]. These results suggested that increased autophagy is an adaptive response to protect chondrocytes from stress and that autophagy regulates OA-like gene expression changes via the modulation of apoptosis and ROS [19]. Thus, the modulators of autophagy, such as mTOR, may represent key targets to for the treatment of OA.

mTOR is a serine/threonine protein kinase that is a negative regulator of autophagy, integrates inputs from nutrients and growth factors, and regulates many basic cellular processes through two distinct protein complexes, mTORC1 and mTORC2. mTOR is an established longevity axis, and its inhibition either pharmacologically or genetically, has been demonstrated to extend lifespan in numerous species [20–22]. Recent studies suggest that mTOR plays an important role in cartilage growth and development, alters articular cartilage homeostasis, and contributes significantly to the cartilage degenerative process associated with OA [23] mTOR expression has been shown to be elevated in OA models and has been associated with increased chondrocyte apoptosis [24]. Given that mTOR is a negative regulator of autophagy, the deleterious effects of age-associated increases in mTOR activity can be linked to decreased autophagy in chondrocytes during OA. Pre-clinical experiments in rat OA models have also shown that suppressing the PI3K/AKT/mTOR signaling pathway promotes articular chondrocyte autophagy and alleviates inflammation [25]. Both pharmacological and genetic approaches for inhibiting mTOR signaling have been shown to reduce the severity of OA in pre-clinical animal models [24–27].

2.2 Pre-clinical studies targeting mTOR for the treatment of OA

Rapamycin is the oldest known natural mTOR inhibitor that has been traditionally used clinically as an immunosuppressive agent [28]. Rapamycin acts through binding of FK506-binding proteins and primarily destabilizes mTORC1, but to some extent can prevent the phosphorylation of downstream targets of mTOR1 [29–31]. Since its FDA approval in 1999, Rapamycin has been used by millions of patients. There is a litany of clinical evidence suggesting that Rapamycin is a safe and effective drug with few side effects for which all are reversible [31, 32].

Interestingly, recent studies have shown that Rapamycin acts to activate human chondrocyte autophagy *in vitro* primarily by inhibiting mTOR complex 1 (mTORC1) and suppressing the development of OA-like changes [19]. Furthermore, systemic administration of Rapamycin has been shown to reduce the severity of OA through activation of autophagy in a mouse model [26].

Our group has found that intra-articular injection of Rapamycin was a safe and effective therapeutic delivery method to protect articular cartilage from osteoarthritic changes in a mouse model of OA [27]. On the other hand, deletion of mTOR has been shown to up-regulate autophagy and protect mice from OA in inducible cartilage-specific mTOR knockout mice [24]. Torin 1, a selective ATP-competitive inhibitor of mTOR, which can cause induction of autophagy, is also regarded as a potent inhibitor of both mTORC1 and mTORC2. In a rabbit model of OA, intra-articular injection of Torin 1 was shown to reduce articular cartilage degeneration [33, 34]. Another agent with demonstrated ability to reduce mTOR activity is Metformin, which is FDA-approved for the treatment of type 2 diabetes [35]. Metformin has been shown to activate 5' AMP-activated protein kinase (AMPK), a negative regulator of mTOR. Recently, Metformin has also been shown to inhibit cartilage degeneration in OA mouse models by downregulating mTOR [36]. In another murine study, Metformin was shown to reduce OA structural worsening and reduce pain scores [37]. Therefore, multiple lines of evidence support the theory of mTOR signaling as a promising therapeutic target for OA, mainly through the modulation of autophagy.

2.3 Autophagy and mTOR in OP

Multiple proteins involved in the autophagic activity are essential for the survival, differentiation, and function of bone cells, including osteocytes, osteoblasts, and osteoclasts [38]. Autophagy is critical for the necessary crosstalk between bone resident cells and thus plays a critical role in the signaling dynamics for bone synthesis (osteoblasts) and degradation (osteoclasts). Dysregulation in the level of autophagic activity has been found to disrupt the balance between bone formation and bone resorption linked to the onset and progression of OP [38–42]. In addition, autophagy plays an important role in MSC function and lineage determination from adipogenesis to osteoblastogenesis, [43] and it has been linked to the increased adipogenic differentiation in bone MSCs [44]. Autophagic activity is known to decrease with age, especially in bone cells, [45] hence the regulation of autophagic activity is considered a promising strategy for the prevention and treatment of OP [14, 46].

Osteoblast dysfunction is a significant cause of aging-related bone loss, but the mechanisms underlying osteoblast dysfunction with aging are not fully elucidated. mTOR has been shown, through in-vitro studies, to regulate osteogenic genes (Runx2, Osterix), stemness genes (Oct3/4, Nanog), and mineralization through alkaline phosphatase production [47].

2.4 Pre-clinical studies targeting mTOR for the treatment of OP

Several studies have investigated the efficacy of targeting mTOR for the treatment of OP. Systemic delivery of autophagy modulators such as Rapamycin and its analogs have been tested in a number of animal models. In a study using 24-month-old rats, micro-CT showed that Rapamycin effectively inhibited aging-related bone loss in trabecular bone. In this study, Rapamycin treatment resulted in a significant decrease in the number of osteoclasts, as well as the induction of osteoclast autophagy and a decrease in osteocyte apoptosis compared to the control group [48]. Besides, mTORC2 signaling stimulates osteoblast differentiation and is involved in aging-related OP [49]. The expression of Rictor, a specific component of mTORC2, is decreased in osteoblasts during aging, which may contribute to aging-related bone loss, and deletion of Rictor in osteoblasts has been shown to accelerate aging-related bone loss in a mouse model [49]. The use of Everolimus, a Rapamycin analog and predominant mTORC1 inhibitor, has been shown to protect against OP onset in ovariectomized rats through the reduction of osteoclast formation and cathepsin K mediated matrix degradation [50]. Rapamycin has also been shown to reduce the severity of age-related bone conditions in trabecular bones of aged male rats by activating osteocyte autophagy [48]. Taken together, mTOR may play a critical role in aging-related OP and represents a promising therapeutic target.

3. Cellular senescence: a new therapeutic strategy for the treatment of aging-related musculoskeletal decline

3.1 Senescent cells

Senescent cells play a vital role in the aging process and promote degenerative diseases, geriatric syndromes, and potentially malignancy through the production of a Senescence Associated Secretory Phenotype (SASP) characterized by pro-inflammatory and catabolic anti-regenerative factors [51]. Senescence is a state defined by replicative arrest and resistance to apoptosis with altered metabolic activity, and several factors, often related to cell or tissue damage, can induce senescence, including DNA lesions, mechanical/shear stress, reactive metabolites, proteotoxic stress, and inflammation [51]. When present, these factors can activate one or more pathways through the p16^{Ink4a}/retinoblastoma protein, p53/p21^{Cip1}, or other transcription factor cascades, resulting in cell cycle arrest, metabolic shifts, altered gene expression, and the production of deleterious SASP factors [51, 52].

Senescent cells accumulate in tissues throughout the lifespan and are normally removed by the immune system [53]. However, inefficient removal due to chronic stress or pathology may exceed the capacity of the immune system, especially during aging, where chronic inflammation disrupts the homeostatic immunologic clearance mechanisms [51, 53–55]. SASP factors from senescent cells are especially deleterious and include various cytokines, chemokines proteins, growth factors, and tissue degrading matrix metalloproteinases (MMPs) [51, 56, 57]. These factors stimulate inflammation, ECM degradation, fibrosis, and secondary senescence in surrounding cells [51, 54, 58]. SASP components can be cell-type specific, and senescence triggers are influenced by several factors, including hormones, stress, drugs, and pathogens [51]. A schema showing the association between senescence and OA is shown in **Figure 1**. With an increased number of senescent cells, a higher amount of secreted SASP components may cause an inflammatory, apoptotic, and cell- and tissue-destructing effect, eventually resulting in aging and chronic diseases. For these reasons, targeting senescent cells has garnered significant attention for the treatment of age-associated pathologies, especially musculoskeletal conditions [59].

According to the "geroscience hypothesis" [60], targeting senescent cells is appealing as they are a fundamental property of aging and may thus delay or reverse physiological consequences of the aging process or the development of aging-related diseases [51, 58]. However, there are no established markers

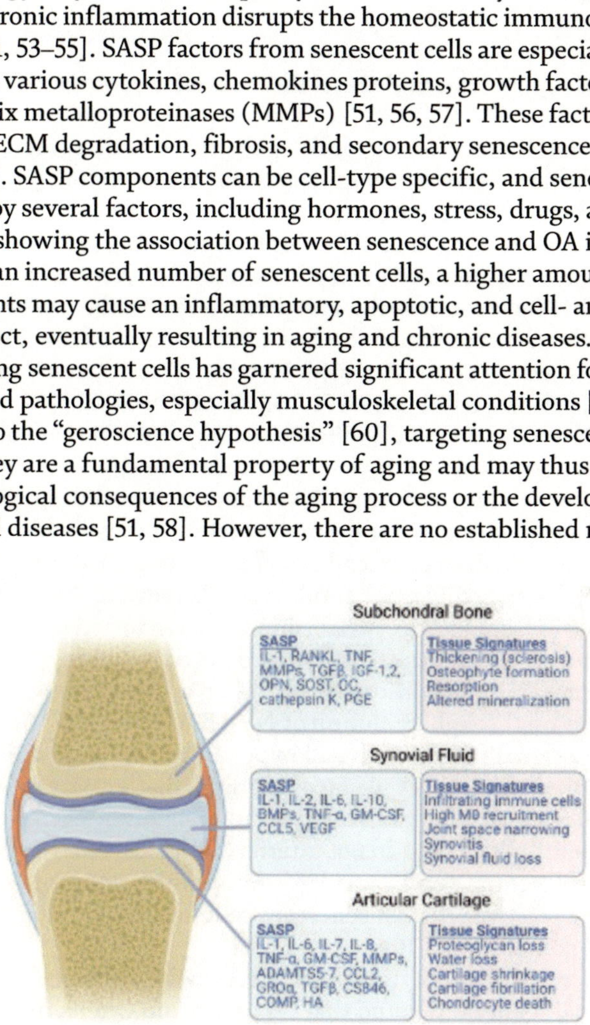

Figure 1.
Schema of the association between senescence and OA. Aging is accompanied by the secretion of the senescence-associated secretory phenotype (SASP), including various chemokines, cytokines, proteases, and growth factors, which act alone or together to cause degenerative changes in the subchondral bone, synovial fold, and articular cartilage, ultimately leading to OA. IL, interleukin; RANKL, Receptor activator of NF-κβ ligand; TNF, tumor necrosis factor; MMPs, Matrix metalloproteinases; TGFβ, Transforming Growth Factor-β; IGF, insulin-like growth factor; OPN, Osteopontin; SOST, Sclerostin; OC, Osteocalcin; PGE, Prostaglandin E; BMPs, Bone morphogenetic proteins; GM-CSF, Granulocyte Macrophage colony-stimulating Factor; CCL, C-C motif chemokine ligand; VEGF, vascular endothelial growth factor; GMCSF, Granulocyte-macrophage-colonystimulating factor; ADAMTS, A disintegrin and metalloproteinase with thrombospondin motifs; GROα, Growth-related oncogene-α; CS-846, The 846 epitope of chondroitin sulfate; COMP, Cartilage oligomeric matrix protein; HA, Hyaluronic acid.

universally specific to senescent cells [51, 54]. Higher expression of $p16^{Ink4a}$ and $p21^{Cip1}$ usually occur in senescent cells [61, 62]. One of the prominent hallmarks, however, is resistance to apoptosis [63]. Zhu et al. reported that senescent cells anti apoptotic pathway (SCAP) networks are expressed at higher levels compared to non-senescent cells, playing an essential role in the resistance to apoptosis [64]. Of 39 transcripts targeted by small interfering RNAs (siRNAs), six transcripts (ephrin ligand (EFN) B1, EFNB3, $p21^{Cip1}$, plasminogen-activated inhibitor-2 (PAI-2), phosphatidylinositol-4,5-bisphosphate 3-kinase delta catalytic subunit (PI3KCD), and BCL-xL) were found to downregulate SCAPs and elicit apoptosis in senescent cells. Targeting several SCAP pathways may also be necessary for senescent cell removal and may further increase specificity for senescent cells while not affecting healthy non-senescent cells [51, 64].

3.2 Treatment of aging-related skeletal diseases with senolytic agents

The use of senolytic agents is a promising approach to delay aging and reduce the severity of chronic diseases through senescent cell depletion [58]. Various drugs have now been characterized that demonstrate the ability to eliminate senescent cells, namely through reduction in anti-apoptotic signaling [65]. For example, Fisetin is a recently characterized senolytic flavonoid phytonutrient found in fruits and vegetables that can extend health and lifespan in naturally aged and progeroid mice [66]. Yousefzadeh et al. tested ten flavonoids and found that Fisetin demonstrated the highest senolytic activity [66]. Fisetin has been found to target senescence associated pathways such as SIRT1, [67] BCL-2/BCL-X_L, [68, 69] HIF-1α, [70] p53/MDM2, [69, 71] and AKT, [69, 72] leading to elimination of senescent cells. However, it also has the ability to reduce oxidative stress via SIRT1/Nrf2, [73] decrease mitochondria-derived ROS via inhibition of GSK3β, [73] and exhibit anti-inflammatory effects. Another example, Metformin, is a widely prescribed anti-diabetic drug with lifespan extending effects in mice, [74] and exhibits senolytic activity via targeting some similar, but also several different senescence associated pathways from Fisetin. Metformin is known to activate AMPK, which indirectly inhibits mTOR, the most established nutrient sensing longevity regulating pathway to date [75]. In addition, Metformin reduces oxidative stress and SASP related inflammation through inhibition of NF-kB activity, [75, 76] and inhibits insulin and IGF-1 signaling, which are all hyperactivated during aging [75, 77]. Thus, many senolytic drugs have pleiotropic effects that mitigate age-related cellular and physiological dysfunction. Furthermore, senolytic agents can be administered intermittently since they do not interfere with a receptor or an enzyme, thereby reducing potential side effects [54]. Accordingly, intermittent treatment may maintain the positive effect of senescent cells on wound healing, cancer prevention, and homeostasis [78–81].

The list of agents with senolytic potential is growing. Zhu et al. tested 46 potential senolytic agents and found that Dasatinib and Quercetin demonstrated particularly good results in terms of reducing senescent cells [64]. Dasatinib is a tyrosine kinase inhibitor used in cancer treatment and is known to interfere with ephrin-dependent suppression of apoptosis [82, 83]. Quercetin is a naturally occurring flavonoid found in several fruits and vegetables known to inhibit PI3K, kinases, and serpines [64]. Noteworthy, Dasatinib, in combination with Quercetin was able to induce apoptosis more effectively in a broader range of senescent cell types than either one alone [84]. Navitoclax, a lymphoma drug, inhibits the anti-apoptotic proteins in the Bcl-2 family and has been used in cancer treatment [85]. It has demonstrated senolytic effects in several cell types *in vitro*. However, in aged mice, it demonstrated reduced trabecular bone volume and impaired osteoblast function [86]. Additionally, Navitoclax is directed against a small number of SCAP network molecules and has shown poor specificity for senescent cells with increased risk of side effects, including thrombocytopenia due to its effect on platelets [87]. Thus, not all drugs with senolytic ability are clinically suitable due to a range of potential side effects or drug-to-drug interactions with existing medications. However, the ability to transiently dose and the existence of natural compounds with favorable safety profiles (ex. Fisetin, Quercetin, Metformin, Rapamycin, Geldanamycin etc.) offer the potential for continuous administration.

Some senolytic agents have been tested for their effects on the musculoskeletal system, demonstrating a potential to reduce aging-related diseases and conditions. Here we will focus on the use of senolytic agents in the setting of OA and OP.

3.2.1 Agents for the treatment of OA

3.2.1.1 Pre-clinical studies of senolytic agents for the treatment of OA

Senescent cells have long been associated with OA [88, 89]. Articular cartilage from patients with advanced OA have a significant number of senescent chondrocytes, [90, 91] with canonical hallmarks of senescence, including metabolic dysfunction, telomere attrition, and decreased autophagy [51, 52]. Aging-related oxidative stress, in addition to mechanical stress from cartilage loading, may contribute to chondrocyte senescence [88]. The cell's ability to compensate for stress reduces with aging, and an accumulation of senescent cells results in a loss of homeostasis with increased secretion of inflammatory mediators, reduced matrix synthesis, and impaired response to growth factor stimulation. This may result in the development of OA. The link between senescence and OA was demonstrated by Xu et al. [92]. Three months after injection of senescent cells into the knee joint of mice, they found changes suggestive of OA, including damage to cartilage and menisci, osteophytes, and changes of the subchondral bone. The mice also demonstrated pain and reduced function. These changes were not found in the control group, suggesting that high concentrations of senescent cells may act detrimentally on the articular cartilage homeostasis [92]. In addition, other groups have shown that local clearance of senescent cells genetically within the intra-articular space significantly reduced the development of injury-induced OA and promoted a pro-regenerative environment [93].

The role of the senescence marker $p16^{Ink4a}$ was studied by Diekman et al. [94]. The mRNA expression of $p16^{Ink4a}$ was significantly higher in aged mice than in young mice, which inhibited chondrocyte proliferation. Interestingly, SASP factors correlated to the expression of $p16^{Ink4a}$ regardless of age; however, inhibition of $p16^{Ink4a}$ did not affect the SASP production or prevent the development of age-induced or post-traumatic OA of the knee joint. In another study, Zheng et al. studied the effect of Fisetin *in vitro* on IL-1β stimulated human chondrocytes and *in vivo* in murine OA models [95]. In the IL-1β stimulated human chondrocytes, Fisetin increased the expression of the enzyme silent information regulator (SIRT) 1 and thereby inhibited the IL-1β induced increased levels of NO, PGE2, IL-6, TNF-α. Additionally, the mRNA expression and the protein levels of the IL-1β induced iNOs, COX-2, MMP-3, MMP-13, ADAMTS-5 were inhibited, and the induced downregulation of SOX-9, aggrecan, and collagen-II degradation was increased. Sirtinol, an inhibitor of SIRT1, reversed the effects of Fisetin on chondrocytes. In the murine OA models, Fisetin demonstrated findings of attenuated progression of OA, thus highlighting its potential as a senolytic treatment for OA [67].

SIRT1 has been reported to be important for cartilage homeostasis by promoting chondrocyte survival and ECM homeostasis [96]. However, data suggest that SIRT1 is proteolytically inactivated during OA [97]. Batshon et al. demonstrated that the NT/CT SIRT1 fragments were found in serum, and an elevated serum NT/CT SIRT1 ratio was associated with both post-traumatic and aging-related OA in mice [98]. A similar increase was found in humans with OA. Further analysis confirmed that the elevated NT/CT SIRT1 fragments are derived from chondrocytes. Senolytic treatment decreased the serum NT/CT SIRT1 ratio and enhanced the intracellular level of SIRT1 in chondrocytes, which correlated with the reduced severity of OA. Dai et al. showed that the combination of Dasatinib and Quercetin to remove senescent chondroprogenitor cells can inhibit SASP formation and thus effectively improve the results of distraction arthroplasty in vitro and in vivo [84]. Recently, it has also been found that intra-articular injection of Navitoclax in post-traumatic OA rats can reduce inflammation, remove senescent chondrocytes in OA, and promote chondrogenic phenotype [99]. Jeon et al. recently tested UBX0101, a senolytic that was found to selectively eliminate senescent cells, found that intra-articular administration of UBX0101 reduced the incidence of post-traumatic OA and associated pain resulting in the development of a prochondrogenic environment [93]. Faust et al. reported that IL-17 expression was increased in mice with post-traumatic OA and in aged mice, and that there was a correlation between senescent cells and IL-17. In their study, senescent fibroblasts increased the level of Th17 cells, when stimulated by IL-6, IL-1β, and TGF-β, and Th17 cells induced senescence in fibroblasts. Inhibition of senescent cells in mice reduced Th17 cells and IL-17; however, both local (UBX0101) and systemic (Navitoclax) senolytic treatment was necessary to reduce cartilage degeneration in aged mice [100]. These findings would provide markers for diagnostic screening and targets for senolytic agents in the treatment of OA.

3.2.1.2 Clinical studies of senolytic agents for the treatment of OA

Several Phase 1 and 2 clinical trials on senolytic agents for the treatment of OA are underway (ClinicalTrials.gov ID: NCT03513016, NCT04229225, NCT04129944, NCT04210986). At our clinic, we are carrying out multiple Phase I/II randomized controlled trials examining the efficacy of Fisetin in the setting of knee OA with (NCT04210986) or without co-treatment with bone marrow concentrate (BMC). In these studies, patients diagnosed with Kellgren-Lawrence grade II-IV knee OA and a numerical rating scale (NRS) pain score of 4–10 are included. Outcome measures include safety and tolerability of Fisetin administration for two days on and 28 days off (20 mg/kg), as well as patient report pain and function indices, OA and SASP related biomarkers, and magnetic resonance imaging (MRI) of cartilage. There have also been multiple other trials using the senolytic agent UBX0101 carried out by Unity Biotechnology (NCT04229225, NCT04129944, NCT03513016, NCT03100799). Many of these studies demonstrated the safety and tolerability of UBX0101 injected intra-articularly with different dosing regimens (single- vs. multi-dose, ascending doses) (https://doi.org/10.1016/j.joca.2020.02.752). However, none of the studies demonstrated a significant reduction in pain up to 12 weeks as assessed by the Western Ontario and McMaster Universities Osteoarthritis Index (WOMAC) scoring system. A long-term outcome trial measuring safety and tolerability of UBX0101 at one year was also terminated, failing to meet primary or secondary objectives. (https://clinicaltrials.gov/ct2/show/NCT04349956) While many of these trials are underway, senolytic treatment for OA may nonetheless be a potentially groundbreaking novel treatment strategy to ameliorate the onset and/or progression of OA. More studies are needed to better understand therapeutic delivery (oral vs.

intra-articular), dosing, and senolytic drug of choice. It stands to reason that a combination of senolytic agents or alternative senolytics with higher potency to eliminate senescent cells of the joint may provide a more effective intervention.

3.2.2 Senolytic agents for the treatment of OP

3.2.2.1 Pre-clinical studies of senolytic agents for the treatment of OP

Several studies have located senescent cells in bone with aging [101, 102]. Farr et al. reported a higher expression of the senolytic markers $p16^{Ink4a}$, $p21^{Cip1}$, and p53, especially in osteocytes and myeloid cells, in aged mice compared to young mice [101]. Corresponding with the senescent osteocytes, the aged mice also demonstrated higher levels of SASP genes. Similar results were reported in bone biopsies from humans [103], suggesting that senescent osteocytes and SASPs may play an important role in aging-related OP. The role of senescent cells was further linked using a transgenic mouse line carrying a suicide transgene (INK-ATTAC) whereby $p16^{Ink4a}$ cells could be eliminated with the treatment of a drug (AP20187). The results showed that the AP20187 treated aged mice demonstrated clearance of senescent cells, lower osteoclast numbers, and improved trabecular bone of the spine and femur compared to the vehicle treated mice. In contrast, treatment with AP20187 in young mice did not change the bone quality. Further, this study showed similar results with oral senolytic treatment using Dasatinib and Quercetin, which led to significantly lower $p16^{Ink4a}$ mRNA expression and percentage of senescent osteocytes in bone compared to vehicle. They also found that the use of the JAK inhibitor Ruxolitinib reduced the SASP factors IL-6, IL-8, and PAI-1 with concomitant improved spine and femur bone microarchitecture [104]. These results suggest that targeting senescent cells or SASP from senescent cells may reduce bone resorption and maintain or enhance bone formation. Therefore, senolytic drugs may be a promising alternative for treating aging-related OP.

3.2.2.2 Clinical studies of senolytic agents for the treatment of OP

Osteoporosis is a debilitating disease that significantly increases the risk of fracture, costing an estimated 13.8 billion USD annually, and directly increases the mortality rate by more than 30% in elderly Americans [105]. Pharmacological treatment for OP consists of two main categories; antiresorptive (bisphosphonates, estrogen agonists, etc.) and anabolic drugs, all with the intent to reduce fracture incidence [106]. However, current therapies are limited given the widely known side effects of chronic use, including the functional decline of the gastrointestinal tract and kidneys, osteonecrosis, esophageal cancer, osteogenic sarcoma, atrial fibrillation, and venous thromboembolism [107]. Further, anti-resorptive therapies, the general first-line approach, are uniformly associated with a concomitant reduction in bone formation, which prevents optimal fracture healing [108]. Thus, disease modifying alternatives with a better safety profile (or that require less dosing) is certainly needed for the treatment of OP. Like OA, cellular senescence is thought to be a fundamental driver of age-associated decline in a bone [104]. Several clinical trials are investigating the use of senotherapeutic drugs in the setting of OP.

In one Phase 2 randomized controlled trial, the senolytic drugs Dasatinib plus Quercetin and Fisetin alone are being tested in healthy elderly women aged 70+, when bone density is known to be reduced (ClinicalTrials.gov ID: NCT04313634). Bone turnover serum markers CTX-I and P2NP are measured with and without senolytic treatment. There are also two other active trials examining the effects of Fisetin for the treatment of frailty syndrome (NCT03675724, NCT03430037), an

		Preclinical Animal Model			Clinical Trial						
Senolytic Agent	Mechanism of Action	Species	Dose	Route	Condition	ID (Phase)	Dose	Route	Target Population	Primary Endpoint	References
Dasatinib (D)	BCR-ABL, SRC, c-KIT, ephrin A receptor, P53 and PAI-2	rat	Mixed solution (1 ml) of D (500 nM) + Q (100 μM, weekly injection until 4 weeks post-op.	IA	OA	NCT04313634 Phase II	1. D + Q treatment group: Intermittent dosing of D (100 mg × 2 days) + Q (1000 mg × 3 consecutive days) every 28 days, repeated 5 times in total.	oral	elderly women	Percent changes in serum bone turnover markers C-terminal telopeptide of type I collagen and amino-terminal propeptide of type I collagen. (Time Frame: 20 weeks)	[59, 84, 109]
Quercetin (Q)	BCL-2/ $BCL\text{-}X_L$ family, PI3K/ AKT, ROS, MDM2/p53/ p21/serpine (PAI-1&2), HIF-1α						2. F treatment group: Intermittent dosing of F (20 mg/kg/day for 3 consecutive days) every 28 days, repeated 5 times in total.	oral	elderly women		
Fisetin (F)	BCL-2/ $BCL\text{-}X_L$ family, PI3K/ AKT, ROS, MDM2/p53/ p21/serpine (PAI-1&2), HIF-1α, SIRT1, IL-1β	mice	F: 20 mg/kg daily	oral	OA	NCT04210986 Phase I/II	F; 20 mg/kg for two consecutive days, followed by 28 days off, then 2 more consecutive days,	oral	Knee OA	Evaluation of liver and kidney toxicity and Tumor Lysis Syndrome by measuring peripheral blood chemistry. (Time Frame: 12 months)	[59, 95, 109]

		Preclinical Animal Model			Clinical Trial						
Senolytic Agent	**Mechanism of Action**	**Species**	**Dose**	**Route**	**Condition**	**ID (Phase)**	**Dose**	**Route**	**Target Population**	**Primary Endpoint**	**References**
Navitoclax (N) (ABT-263)	BCL-2/BCL-X_L family	rat	N: 0 25–5 μM, 4 injections within 4 to 6 weeks post-op.	IA	OA	N/A	N/A	N/A	N/A	N/A	[59, 99, 109]
UBX0101 (U)	MDM2/p53	mice	U: 1 mM, every other day starting from 2 weeks post-op. (5–6 shots)	IA	OA	NCT03513016 Phase I	U: 0.1–0 4 mg, single dose	IA	Knee OA	Safety and tolerability of a single dose of U. (Time Frame: 12 weeks)	[59, 93, 109]
						NCT04229225 Phase I	U: Single injection of 8.0 mg at week 0, or two doses of 4.0 mg (weeks 0 and 4).	IA	Knee OA	Safety and tolerability of a single and repeat dose of U. (Time Frame: 24 weeks)	
						NCT04129944 Phase II	U: Single dose of 0.5 mg, 2 mg, or 4 mg at week 0.	IA	Knee OA	Change in (WOMAC-A) Score from baseline to week 12. (Time Frame: 12 weeks)	

IA, intra-articular injection.

Table 1.
Senolytics as Potential Therapeutic Agents for OA and OP.

age-related condition characterized by sarcopenia and decreased bone density. In these studies, primary endpoints include serum inflammatory markers and mobility based on a 6-min walk test.

Another promising senolytic drug is Metformin. There are at least four active trials measuring the effects of Metformin on pre-frail to frail patients (NCT03451006, NCT02570672, NCT04221750, NCT02325245). Primary endpoints in these studies are mobility and motor skills functions, including gait speed, balance ability, and grip strength test, geriatric depression score, and weight loss. Similar to OA trials, many of these trials are not complete, but there is compelling evidence that senolytic agents might benefit a litany of age-related skeletal decline. Details of preclinical animal studies and clinical trials of major senolytics in OA and OP and their mechanisms are summarized in **Table 1** [59, 109].

4. Mesenchymal stem cell (MSC)-based therapy for the treatment of aging-related musculoskeletal decline

4.1 Biological mechanisms of MSCs

Mesenchymal stem cells (MSCs) are present in a variety of human tissues, including bone marrow, adipose tissue, synovial tissue, and cord blood [110–112]. However, there is currently a lack of conclusive evidence regarding the potential biological mechanisms of MSCs for the treatment of aging-related musculoskeletal diseases. Understanding the function of MSCs opens up the possibility of developing robust MSC-based therapies for musculoskeletal regenerative medicine. Until now, there are two theories on the mechanism of function: (1) Direct adherence and incorporation into the host tissue and (2) trophic effects resulting from the MSC-derived secretomes.

4.1.1 Direct adherence and incorporation of MSCs

One of the primary advantages of MSCs is their ability to interact with various chemokine receptors (such as CXCR4, which is involved in MSC migration), integrins, selectins, and vascular cell adhesion molecule-1, to home damaged tissues [113–119]. The original hypothesis regarding the tissue regeneration mechanism of MSCs was that implanted cells would migrate directly to injury sites, where they would differentiate into functional cells and eventually promote repair of damaged connective tissue [120]. In support of this hypothesis, it has been reported that injected MSCs have the potential to adhere to the injury site and repair the host cartilage through regeneration, and interestingly, MSCs may also migrate to the injury site for tissue regeneration [121, 122]. However, whether the introduced MSCs are actually taken up into the host tissue and act directly on the damaged tissue has yet to be verified.

4.1.2 Trophic effects of MSCs

With decades of research on the underlying functionality of MSCs, it has been found that there exists a discrepancy between the frequency and duration of transplants and the remarkable healing power of MSCs [120]. Numerous studies have been conducted to resolve this conundrum, presenting the concept that MSCs possess the ability to maintain the proliferation and survival of certain cell types by secreting trophic factors, and regulating certain aspects of the immune system, thereby ushering MSC-based therapies into a new phase [123].

Analysis of the MSC secretion and proteome has revealed various paracrine factors that can reduce apoptosis and inflammation and stimulate angiogenesis and self-renewal of progenitor cells [120]. Notably, MSCs are known to act as immunosuppressive cells that can alleviate inflammation and reduce monocyte activation by releasing anti-inflammatory factors, including interleukin-1 receptor antagonist (IL-1ra) [124]. Pro-inflammatory cytokines, such as interleukin-1β (IL-1β), are widely known to play an essential role in OA progression [95] and IL-1ra confers an overall inhibitory effect on IL-1ß mediated inflammation and matrix degradation. Taken together, MSCs should confer a therapeutic potential in OA patients.

4.2 Properties and benefits of each type of MSCs on aging-related musculoskeletal decline

4.2.1 Bone marrow-derived MSCs (BM-MSCs)

4.2.1.1 Characteristics and advantages of BM-MSCs

Bone marrow is generally considered to be the home of hematopoietic stem cells and is known to contain MSCs as part of the stromal fraction [125]. BM-MSCs possess a high potential for cartilage repair due to their ready availability [126]. BM-MSCs have also been widely studied as a treatment for OP due to their high ability of osteogenesis.

4.2.1.2 Pre-clinical studies of BM-MSCs for OA

The effect of BM-MSCs on OA has been verified in numerous animal studies. Chiang et al. investigated the effects of intra-articular injection of allogeneic BM-MSCs in an *in vivo* rabbit OA model. They observed that the BM-MSCs transplantation group had significantly better histological scores than the hyaluronic acid injection group [127]. Furthermore, Song et al. compared the efficacy of bone marrow mononuclear cells (BMMCs) and BM-MSCs in a sheep OA model and demonstrated that the BM-MSCs group had smaller lesions and a relatively smoother femoral condyle. They also reported that ICRS scores showed a greater improvement in the BM-MSCs group than the BMMCs and PBS (control) groups. They further stated that the results of histology showed fewer changes to cartilage and bone in the BM-MSCs group [128].

4.2.1.3 Pre-clinical studies of bm-MSCs for OP

Ichioka et al. demonstrated that direct injection of allogenic BM-MSCs into the bone marrow cavity of irradiated P6 sub-strain of senescence-accelerated mice (SAMP6), an osteoporotic mouse, resulted in inhibition of osteoblast and osteoclast formation in an age-dependent manner and promoted adipogenesis, increased trabecular bone and decreased bone mineral density [129]. Autologous BM-MSC transplantation has been reported to improve bone formation and strengthen osteoporotic bones in ovariectomy (OVX) -treated rabbits [130] and in estrogen-deficient goats, mimicking the postmenopausal OP that occurs in elderly women [110]. However, there is limited support for autologous BM-MSCs to treat OP in elderly patients because BM-MSCs isolated from the bone marrow of elderly patients have shown reduced proliferation and osteogenic capacity in vitro [131, 132].

4.2.2 Adipose-derived MSCs (A-MSCs)

4.2.2.1 Characteristics and advantages of A-MSCs

MSCs were first reported in adipose tissue in 2001 [133] and have been touted as an attractive source of MSCs. Although A-MSCs have the advantage of being easier to isolate than BM-MSCs, [134] BM-MSCs have been shown to be prone to chondrogenic differentiation, both *in vitro* and *in vivo* [135]. Interestingly, however, it has also been reported that the addition of paracrine or cytokine factors increases the cartilage capacity of A-MSCs to a level comparable to that of BM-MSCs [136]. It is worth noting that the yield of A-MSCs and their proliferation and differentiation ability is dependent on the site of tissue collection [137] and the age of the donor [138].

4.2.2.2 Pre-clinical studies of A-MSCs for OA

The effect of A-MSCs on OA has been investigated in numerous animal studies. Tang et al. compared the efficacy of three types of intra-articular injections, subcutaneous A-MSCs, visceral A-MSCs, and PBS (control), in a rat model of OA. Subcutaneous They found that A-MSCs injection decreased osteophyte and fibrous tissue formation compared to PBS or visceral A-MSCs. In addition, histologically, a smooth cartilage surface and distribution of lacunae and chondrocytes were observed in rats treated with subcutaneous A-MSCs [139]. Kuroda et al. verified the efficacy of A-MSCs for OA treatment using a rabbit model. They found that nearly normal cartilage was observed in the A-MSCs group, with less cartilage damage than in the control group. Further, the proportion of MMP-13-positive cells was significantly lower in sections of the A-MSCs group than in the control group [140].

4.2.2.3 Pre-clinical studies of A-MSCs for OP

The effects of A-MSCs have also been evaluated in OP animal models. Mirsaidi et al. performed the A-MSCs injection to SAMP6 mice and observed improvement of trabecular bone quality [141]. Additionally, Cho et al. studied the efficacy of human A-MSCs using OVX nude mice, showing that human A-MSCs could inhibit OVX-induced bone loss over eight weeks [142]. Furthermore, Ye et al. found that autologous A-MSCs enhanced bone regeneration in an OVX-induced rabbit model of OP, suggesting that this was due not only to autologous osteogenic differentiation but also to the promotion of osteogenesis and inhibition of adipogenesis through the activation of BMP-2 and BMPR-IB signaling pathways [143].

4.2.3 Synovium-derived MSCs (S-MSCs)

4.2.3.1 Characteristics and advantages of S-MSCs

In 2001, De Bari *et al*. isolated the first MSCs from the synovium of the human knee joint [144]. Since then, S-MSCs have attracted attention because they are more readily accessible, possess a higher growth rate, and are less immunogenic compared with MSCs from other origins [145]. Sakaguchi et al. compared the yield, expandability, differentiation potential, and epitope profiles of human MSCs derived from five different mesenchymal tissue sources: bone marrow, synovium, periosteum, adipose tissue, and muscle, and concluded that S-MSCs had the highest capacity for chondrogenesis [146].

4.2.3.2 Pre-clinical studies of S-MSCs for OA

The beneficial effects of S-MSCs in promoting cartilage regeneration have been reported in pig, [146] leporin, [135] and canine models [147]. In a recent systematic review of *in vivo* studies on synovium-derived mesenchymal stem cell transplantation for cartilage regeneration, To et al. showed, in 4 human and 16 animal articles, that S-MSCs possess overall good chondrogenic potential and positive effect for treating chondral lesions and preventing OA [148]. Ozeki et al. found that intra-articular injection of S-MSCs in a rat model of OA could inhibit the OA progression and attenuate synovitis when administered once a week instead of a single dose [149]. Accumulating evidence that S-MSCs possess a strong chondrogenic potential and the fact that MSCs derived from synovial tissue is specific to target joints have led to a growing interest in the application of S-MSCs for a stem cell therapy of knee OA.

4.2.3.3 Pre-clinical studies of S-MSCs for OP

As it pertains to osteogenic potential, Sakaguchi et al. showed that S-MSCs possessed a higher capacity than adipose tissue- and muscle-derived cells, comparable to bone marrow-, and periosteal-derived [150]. However, S-MSC-based therapy for OP has yet to be well investigated.

4.2.4 Muscle-derived stem cells (MDSCs)/muscle-derived stem progenitor cells (MDSPCs)

4.2.4.1 Characteristics and advantages of MDSCs/MDSPCs

MDSCs/MDSPCs are pluripotent cells isolated from postnatal skeletal muscle via established preplating techniques. They are characterized by multiple critical features such as long-term proliferation/self-renewing capacity, resistance to oxidative and inflammatory stress, and multilineage differentiation potential [151–153]. Recently, it has been shown that skeletal muscle-derived MSCs from OA patients exhibit superior biological properties compared to the bone-derived MSCs counterpart, making them a promising candidate for autologous stem cell therapy [154]. MDSPCs have been shown to improve the regenerative capacity of various tissues, including bone, cartilage, skeletal muscle, and cardiac muscle, by promoting angiogenesis [155–159].

Of note, several studies have investigated differences in the proliferation and differentiation ability of MDSCs by sex and age [158, 160, 161]. It has been shown that male murine MDSPCs displayed higher chondrogenic differentiation capacity and cartilage regeneration potential than female murine MDSPCs [160]. Similarly, Corsi et al. showed that the osteogenesis of male murine MDSPCs was superior to that of female MDSPCs [162]. Furthermore, our group has recently found that in human MDSPCs, male MDSPCs possess a greater ability to undergo chondrogenesis and osteogenesis than female MDSPCs [161].

Our group, on the other hand, have reported that not only donor but also host sex affects bone regeneration; male murine hosts showed a greater amount of MDSPC-mediated ectopic bone formation and cranial defect healing than did female hosts [163].

4.2.4.2 Pre-clinical studies of MDSCs/MDSPCs for OA

An interesting study by Kuroda et al. indicated that local delivery of BMP-4 by genetically engineered MDSCs promoted chondrogenesis with a significant improvement of articular cartilage repair in rats [164]. This suggests that MDSCs are advantageous concerning chondrogenic differentiation potential. Furthermore, Matsumoto et al. demonstrated in a rat model that MDSCs therapy with sFlt-1 and BMP-4 promotes chondrogenesis in OA, and inhibits cartilage resorption by inhibiting angiogenesis, thus enabling sustained cartilage regeneration and repair [165].

4.2.4.3 Pre-clinical studies of MDSCs/MDSPCs for OP

Our group isolated young and old populations of gender-matched human muscle-derived stem cells (hMDSCs) to examine the effect of age on osteogenic differentiation using a critical-size calvarial bone defect mouse model. In addition, the effect of donor and host age on hMDSC-mediated bone regeneration was investigated. We showed that donor age did not impair hMDSC-mediated bone regeneration, while host age had the adverse effect. We also found that hMDSCs form functional bone regardless of the age of the donor or host, suggesting that these cells are a promising resource for bone regeneration [166].

4.3 Clinical studies of MSC-based therapy for the treatment of aging-related musculoskeletal decline

4.3.1 Clinical studies of MSC-based therapy for OA

More than 30 clinical trials on the administration of intra-articular MSCs for the treatment of OA have been completed to date, including randomized controlled trials, retrospective studies, and cohort studies (https://www.clinicaltrials.gov/), and most of the published results have shown clinical benefit [111, 112, 167–178]. Currently, B-MSCs and A-MSCs are the most commonly used cell sources in clinical trials for OA, [179] however, finding an optimal treatment with MSCs is challenging due to the great diversity of patient populations, delivery methods, cell numbers, culture expansion methodology, and follow-up periods.

Until now, no clinical trials on the benefit of intra-articular administration of S-MSCs in patients with OA have been published. Interestingly, however, Sekiya et al. found that arthroscopic S-MSCs transplantation improved the clinical outcome of knees with articular cartilage defects at a mean follow-up of 52 months in 10 patients based on MRI, histology, and clinical outcome score evaluation [180]. Furthermore, Shimomura et al. conducted the first human pilot study of implanting scaffold-free tissue-engineered constructs generated from S-MSCs to the injury site for five patients with symptomatic knee cartilage lesions, demonstrating that self-assessed clinical scores for pain, symptoms, activities of daily living, sports activities, and quality of life improved significantly at 24 months postoperatively, with no serious adverse events. In addition, second-look arthroscopy and MRI confirmed complete defect filling in all cases, and biopsy of the regenerated cartilage showed that the repair tissue consisted of hyaline cartilage [181]. These results showed that implantation of S-MSCs could repair articular cartilage damage and prevent its progression to OA, and therefore, future clinical applications in patients with OA may be promising. There is also evidence that, in a phase I/II study, repeated administration of umbilical cord-derived-MSCs improved safety and clinical outcomes for long-term pain in patients with knee OA [112].

4.3.2 Clinical studies of MSC-based therapy for OP

A recent Phase 1 clinical trial has been conducted using fucosylated BM-MSCs for patients with OP. In this study, autologous BM-MSCs were harvested 30 days prior to the infusion and cultured under good manufacturing practice (GMP) conditions to purify and obtain mesenchymal cell established dose range. This study is ongoing, and the recruitment of participants is currently closed. (ClinicalTrials.gov ID: NCT02566655).

Although the use of MSC products for the treatment of aging-related skeletal disorders is becoming increasingly prevalent, well-designed studies are imperative to conclusively prove their clinical benefits and identify the optimal indications, cell sources, delivery methods, and doses.

Author details

Naomasa Fukase, Ingrid K. Stake, Yoichi Murata, William S. Hambright, Sudheer Ravuri, Marc J. Philippon and Johnny Huard*
Steadman Philippon Research Institute, Vail, CO, USA

*Address all correspondence to: jhuard@sprivail.org

References

[1] Lopez-Otin, C., et al., *The hallmarks of aging.* Cell, 2013. **153**(6): p. 1194-217.

[2] Loeser, R.F., et al., *Osteoarthritis: a disease of the joint as an organ.* Arthritis Rheum, 2012. **64**(6): p. 1697-707.

[3] Vignon, E., et al., *Quantitative histological changes in osteoarthritic hip cartilage. Morphometric analysis of 29 osteoarthritic and 26 normal human femoral heads.* Clin Orthop Relat Res, 1974(103): p. 269-78.

[4] Kronenberg, H.M., *Developmental regulation of the growth plate.* Nature, 2003. **423**(6937): p. 332-6.

[5] Hardingham, T. and M. Bayliss, *Proteoglycans of articular cartilage: changes in aging and in joint disease.* Semin Arthritis Rheum, 1990. **20**(3 Suppl 1): p. 12-33.

[6] Dejica, V.M., et al., *Increased type II collagen cleavage by cathepsin K and collagenase activities with aging and osteoarthritis in human articular cartilage.* Arthritis Res Ther, 2012. **14**(3): p. R113.

[7] van der Kraan, P.M. and W.B. van den Berg, *Chondrocyte hypertrophy and osteoarthritis: role in initiation and progression of cartilage degeneration?* Osteoarthritis Cartilage, 2012. **20**(3): p. 223-32.

[8] Osterhoff, G., et al., *Bone mechanical properties and changes with osteoporosis.* Injury, 2016. **47 Suppl 2**: p. S11-20.

[9] Kassem, M. and P.J. Marie, *Senescence-associated intrinsic mechanisms of osteoblast dysfunctions.* Aging Cell, 2011. **10**(2): p. 191-7.

[10] Marie, P.J. and M. Kassem, *Extrinsic mechanisms involved in age-related defective bone formation.* J Clin Endocrinol Metab, 2011. **96**(3): p. 600-9.

[11] Curtis, E., et al., *Determinants of Muscle and Bone Aging.* J Cell Physiol, 2015. **230**(11): p. 2618-25.

[12] Rachner, T.D., S. Khosla, and L.C. Hofbauer, *Osteoporosis: now and the future.* Lancet, 2011. **377**(9773): p. 1276-87.

[13] Yu, B. and C.Y. Wang, *Osteoporosis: The Result of an 'Aged' Bone Micro-environment.* Trends Mol Med, 2016. **22**(8): p. 641-644.

[14] Levine, B. and D.J. Klionsky, *Development by self-digestion: molecular mechanisms and biological functions of autophagy.* Dev Cell, 2004. **6**(4): p. 463-77.

[15] Levine, B. and G. Kroemer, *Autophagy in the pathogenesis of disease.* Cell, 2008. **132**(1): p. 27-42.

[16] Lepetsos, P. and A.G. Papavassiliou, *ROS/oxidative stress signaling in osteoarthritis.* Biochim Biophys Acta, 2016. **1862**(4): p. 576-591.

[17] Carames, B., et al., *Autophagy is a protective mechanism in normal cartilage, and its aging-related loss is linked with cell death and osteoarthritis.* Arthritis Rheum, 2010. **62**(3): p. 791-801.

[18] Wu, C., et al., *Defective autophagy in chondrocytes with Kashin-Beck disease but higher than osteoarthritis.* Osteoarthritis Cartilage, 2014. **22**(11): p. 1936-46.

[19] Sasaki, H., et al., *Autophagy modulates osteoarthritis-related gene expression in human chondrocytes.* Arthritis Rheum, 2012. **64**(6): p. 1920-8.

[20] Vellai, T., et al., *Genetics: influence of TOR kinase on lifespan in C. elegans.* Nature, 2003. **426**(6967): p. 620.

[21] Kapahi, P., et al., *Regulation of lifespan in Drosophila by modulation of genes in the TOR signaling pathway.* Curr Biol, 2004. **14**(10): p. 885-90.

[22] Kaeberlein, M., et al., *Regulation of yeast replicative life span by TOR and Sch9 in response to nutrients.* Science, 2005. **310**(5751): p. 1193-6.

[23] Pal, B., et al., *mTOR: a potential therapeutic target in osteoarthritis?* Drugs R D, 2015. **15**(1): p. 27-36.

[24] Zhang, Y., et al., *Cartilage-specific deletion of mTOR upregulates autophagy and protects mice from osteoarthritis.* Ann Rheum Dis, 2015. **74**(7): p. 1432-40.

[25] Xue, J.F., et al., *Inhibition of PI3K/AKT/mTOR signaling pathway promotes autophagy of articular chondrocytes and attenuates inflammatory response in rats with osteoarthritis.* Biomed Pharmacother, 2017. **89**: p. 1252-1261.

[26] Caramés, B., et al., *Autophagy activation by rapamycin reduces severity of experimental osteoarthritis.* Ann Rheum Dis, 2012. **71**(4): p. 575-81.

[27] Takayama, K., et al., *Local intra-articular injection of rapamycin delays articular cartilage degeneration in a murine model of osteoarthritis.* Arthritis Res Ther, 2014. **16**(6): p. 482.

[28] Hartford, C.M. and M.J. Ratain, *Rapamycin: something old, something new, sometimes borrowed and now renewed.* Clin Pharmacol Ther, 2007. **82**(4): p. 381-8.

[29] Papadopoli, D., et al., *mTOR as a central regulator of lifespan and aging.* F1000Res, 2019. **8**.

[30] Hausch, F., et al., *FKBPs and the Akt/mTOR pathway.* Cell Cycle, 2013. **12**(15): p. 2366-70.

[31] Hambright, W.S., M.J. Philippon, and J. Huard, *Rapamycin for aging stem cells.* Aging (Albany NY), 2020. **12**(15): p. 15184-15185.

[32] Blagosklonny, M.V., *Rapamycin for longevity: opinion article.* Aging (Albany NY), 2019. **11**(19): p. 8048-8067.

[33] Cheng, N.T., A. Guo, and Y.P. Cui, *Intra-articular injection of Torin 1 reduces degeneration of articular cartilage in a rabbit osteoarthritis model.* Bone Joint Res, 2016. **5**(6): p. 218-24.

[34] Liu, Q., et al., *Discovery of 1-(4-(4-propionylpiperazin-1-yl)-3-(trifluoromethyl)phenyl)-9-(quinolin-3-yl)benzo[h][1,6]naphthyridin-2(1H)-one as a highly potent, selective mammalian target of rapamycin (mTOR) inhibitor for the treatment of cancer.* J Med Chem, 2010. **53**(19): p. 7146-55.

[35] Kinaan, M., H. Ding, and C.R. Triggle, *Metformin: An Old Drug for the Treatment of Diabetes but a New Drug for the Protection of the Endothelium.* Med Princ Pract, 2015. **24**(5): p. 401-15.

[36] Feng, X., et al., *Metformin attenuates cartilage degeneration in an experimental osteoarthritis model by regulating AMPK/mTOR.* Aging (Albany NY), 2020. **12**(2): p. 1087-1103.

[37] Li, H., et al., *Exploration of metformin as novel therapy for osteoarthritis: preventing cartilage degeneration and reducing pain behavior.* Arthritis Res Ther, 2020. **22**(1): p. 34.

[38] Yin, X., et al., *Autophagy in bone homeostasis and the onset of osteoporosis.* Bone Res, 2019. **7**: p. 28.

[39] Yin, X., et al., *Correction to: Autophagy in bone homeostasis and the onset of osteoporosis.* Bone Res, 2020. **8**: p. 36.

[40] Shapiro, I.M., et al., *Boning up on autophagy: the role of autophagy in skeletal biology.* Autophagy, 2014. **10**(1): p. 7-19.

[41] Pierrefite-Carle, V., et al., *Autophagy in bone: Self-eating to stay in balance.* Ageing Res Rev, 2015. **24**(Pt B): p. 206-17.

[42] Greenhill, C., *Bone: Autophagy regulates bone growth in mice.* Nat Rev Endocrinol, 2016. **12**(1): p. 4.

[43] Pantovic, A., et al., *Coordinated time-dependent modulation of AMPK/Akt/mTOR signaling and autophagy controls osteogenic differentiation of human mesenchymal stem cells.* Bone, 2013. **52**(1): p. 524-31.

[44] Nuschke, A., et al., *Human mesenchymal stem cells/multipotent stromal cells consume accumulated autophagosomes early in differentiation.* Stem Cell Res Ther, 2014. **5**(6): p. 140.

[45] Hubbard, V.M., et al., *Selective autophagy in the maintenance of cellular homeostasis in aging organisms.* Biogerontology, 2012. **13**(1): p. 21-35.

[46] Kroemer, G., *Autophagy: a druggable process that is deregulated in aging and human disease.* J Clin Invest, 2015. **125**(1): p. 1-4.

[47] Lee, K.W., et al., *Rapamycin promotes the osteoblastic differentiation of human embryonic stem cells by blocking the mTOR pathway and stimulating the BMP/Smad pathway.* Stem Cells Dev, 2010. **19**(4): p. 557-68.

[48] Luo, D., et al., *Rapamycin reduces severity of senile osteoporosis by activating osteocyte autophagy.* Osteoporos Int, 2016. **27**(3): p. 1093-1101.

[49] Lai, P., et al., *Loss of Rictor with aging in osteoblasts promotes age-related bone loss.* Cell Death Dis, 2016. 7(10): p. e2408.

[50] Kneissel, M., et al., *Everolimus suppresses cancellous bone loss, bone resorption, and cathepsin K expression by osteoclasts.* Bone, 2004. **35**(5): p. 1144-56.

[51] Kirkland, J.L. and T. Tchkonia, *Senolytic drugs: from discovery to translation.* J Intern Med, 2020. **288**(5): p. 518-536.

[52] Campisi, J., *Senescent cells, tumor suppression, and organismal aging: good citizens, bad neighbors.* Cell, 2005. **120**(4): p. 513-22.

[53] Prata, L., et al., *Senescent cell clearance by the immune system: Emerging therapeutic opportunities.* Semin Immunol, 2018. **40**: p. 101275.

[54] Wissler Gerdes, E.O., et al., *Discovery, development, and future application of senolytics: theories and predictions.* FEBS J, 2020. **287**(12): p. 2418-2427.

[55] Kaur, J. and J.N. Farr, *Cellular senescence in age-related disorders.* Transl Res, 2020. **226**: p. 96-104.

[56] Coppe, J.P., et al., *Senescence-associated secretory phenotypes reveal cell-nonautonomous functions of oncogenic RAS and the p53 tumor suppressor.* PLoS Biol, 2008. **6**(12): p. 2853-68.

[57] Kuilman, T. and D.S. Peeper, *Senescence-messaging secretome: SMS-ing cellular stress.* Nat Rev Cancer, 2009. **9**(2): p. 81-94.

[58] Martel, J., et al., *Emerging use of senolytics and senomorphics against aging and chronic diseases.* Med Res Rev, 2020. **40**(6): p. 2114-2131.

[59] Kirkland, J.L., et al., *The Clinical Potential of Senolytic Drugs.* J Am Geriatr Soc, 2017. **65**(10): p. 2297-2301.

[60] Kritchevsky, S.B. and J.N. Justice, *Testing the Geroscience Hypothesis: Early Days.* J Gerontol A Biol Sci Med Sci, 2020. **75**(1): p. 99-101.

[61] Frescas, D., et al., *Murine mesenchymal cells that express elevated levels of the CDK inhibitor p16(Ink4a) in vivo are not necessarily senescent.* Cell Cycle, 2017. **16**(16): p. 1526-1533.

[62] Hsu, C.H., S.J. Altschuler, and L.F. Wu, *Patterns of Early p21 Dynamics Determine Proliferation-Senescence Cell Fate after Chemotherapy.* Cell, 2019. **178**(2): p. 361-373.e12.

[63] Hernandez-Segura, A., J. Nehme, and M. Demaria, *Hallmarks of Cellular*

Senescence. Trends Cell Biol, 2018. **28**(6): p. 436-453.

[64] Zhu, Y., et al., *The Achilles' heel of senescent cells: from transcriptome to senolytic drugs.* Aging Cell, 2015. **14**(4): p. 644-58.

[65] Childs, B.G., et al., *Senescent cells: an emerging target for diseases of ageing.* Nat Rev Drug Discov, 2017. **16**(10): p. 718-735.

[66] Yousefzadeh, M.J., et al., *Fisetin is a senotherapeutic that extends health and lifespan.* EBioMedicine, 2018. **36**: p. 18-28.

[67] Zheng, W., et al., *Fisetin inhibits IL-1beta-induced inflammatory response in human osteoarthritis chondrocytes through activating SIRT1 and attenuates the progression of osteoarthritis in mice.* Int Immunopharmacol, 2017. **45**: p. 135-147.

[68] Pal, H.C., et al., *Fisetin inhibits growth, induces G(2)/M arrest and apoptosis of human epidermoid carcinoma A431 cells: role of mitochondrial membrane potential disruption and consequent caspases activation.* Exp Dermatol, 2013. **22**(7): p. 470-5.

[69] Zhu, Y., et al., *New agents that target senescent cells: the flavone, fisetin, and the BCL-XL inhibitors, A1331852 and A1155463.* Aging (Albany NY), 2017. **9**(3): p. 955-963.

[70] Triantafyllou, A., et al., *Flavonoids induce HIF-1alpha but impair its nuclear accumulation and activity.* Free Radic Biol Med, 2008. **44**(4): p. 657-70.

[71] Li, J., et al., *Fisetin, a dietary flavonoid, induces cell cycle arrest and apoptosis through activation of p53 and inhibition of NF-kappa B pathways in bladder cancer cells.* Basic Clin Pharmacol Toxicol, 2011. **108**(2): p. 84-93.

[72] Zhang, X.J. and S.S. Jia, *Fisetin inhibits laryngeal carcinoma through regulation of AKT/NF-kappaB/mTOR and ERK1/2 signaling pathways.* Biomed Pharmacother, 2016. **83**: p. 1164-1174.

[73] Shanmugam, K., et al., *Fisetin Confers Cardioprotection against Myocardial Ischemia Reperfusion Injury by Suppressing Mitochondrial Oxidative Stress and Mitochondrial Dysfunction and Inhibiting Glycogen Synthase Kinase 3beta Activity.* Oxid Med Cell Longev, 2018. **2018**: p. 9173436.

[74] Martin-Montalvo, A., et al., *Metformin improves healthspan and lifespan in mice.* Nat Commun, 2013. **4**: p. 2192.

[75] Barzilai, N., et al., *Metformin as a Tool to Target Aging.* Cell Metab, 2016. **23**(6): p. 1060-1065.

[76] Moiseeva, O., et al., *Metformin inhibits the senescence-associated secretory phenotype by interfering with IKK/NF-kappaB activation.* Aging Cell, 2013. **12**(3): p. 489-98.

[77] Liu, B., et al., *Potent anti-proliferative effects of metformin on trastuzumab-resistant breast cancer cells via inhibition of erbB2/IGF-1 receptor interactions.* Cell Cycle, 2011. **10**(17): p. 2959-66.

[78] Demaria, M., et al., *An essential role for senescent cells in optimal wound healing through secretion of PDGF-AA.* Dev Cell, 2014. **31**(6): p. 722-33.

[79] Munoz-Espin, D. and *M. Serrano*, *Cellular senescence: from physiology to pathology.* Nat Rev Mol Cell Biol, 2014. **15**(7): p. 482-96.

[80] Chiche, A., et al., *Injury-Induced Senescence Enables In Vivo Reprogramming in Skeletal Muscle.* Cell Stem Cell, 2017. **20**(3): p. 407-414 e4.

[81] Rhinn, M., B. Ritschka, and W.M. Keyes, *Cellular senescence in development, regeneration and disease.* Development, 2019. **146**(20).

[82] Montero, J.C., et al., *Inhibition of SRC family kinases and receptor tyrosine kinases by dasatinib: possible combinations in solid tumors.* Clin Cancer Res, 2011. **17**(17): p. 5546-52.

[83] Xi, H.Q., et al., *Eph receptors and ephrins as targets for cancer therapy.* J Cell Mol Med, 2012. **16**(12): p. 2894-909.

[84] Dai, H., et al., *Eliminating senescent chondrogenic progenitor cells enhances chondrogenesis under intermittent hydrostatic pressure for the treatment of OA.* Stem Cell Res Ther, 2020. **11**(1): p. 199.

[85] Zhu, Y., et al., *Identification of a novel senolytic agent, navitoclax, targeting the Bcl-2 family of anti-apoptotic factors.* Aging Cell, 2016. **15**(3): p. 428-35.

[86] Sharma, A.K., et al., *The Senolytic Drug Navitoclax (ABT-263) Causes Trabecular Bone Loss and Impaired Osteoprogenitor Function in Aged Mice.* Front Cell Dev Biol, 2020. **8**: p. 354.

[87] Croce, C.M. and J.C. Reed, *Finally, An Apoptosis-Targeting Therapeutic for Cancer.* Cancer Res, 2016. **76**(20): p. 5914-5920.

[88] Loeser, R.F., *Aging and osteoarthritis: the role of chondrocyte senescence and aging changes in the cartilage matrix.* Osteoarthritis Cartilage, 2009. **17**(8): p. 971-9.

[89] Loeser, R.F., J.A. Collins, and B.O. Diekman, *Ageing and the pathogenesis of osteoarthritis.* Nat Rev Rheumatol, 2016. **12**(7): p. 412-20.

[90] Martin, J.A. and J.A. Buckwalter, *Telomere erosion and senescence in human articular cartilage chondrocytes.* J Gerontol A Biol Sci Med Sci, 2001. **56**(4): p. B172-9.

[91] Harbo, M., et al., *The relationship between ultra-short telomeres, aging of articular cartilage and the development of human hip osteoarthritis.* Mech Ageing Dev, 2013. **134**(9): p. 367-72.

[92] Xu, M., et al., *Transplanted Senescent Cells Induce an Osteoarthritis-Like Condition in Mice.* J Gerontol A Biol Sci Med Sci, 2017. **72**(6): p. 780-785.

[93] Jeon, O.H., et al., *Local clearance of senescent cells attenuates the development of post-traumatic osteoarthritis and creates a pro-regenerative environment.* Nat Med, 2017. **23**(6): p. 775-781.

[94] Diekman, B.O., et al., *Expression of p16(INK) (4a) is a biomarker of chondrocyte aging but does not cause osteoarthritis.* Aging Cell, 2018. **17**(4): p. e12771.

[95] Zheng, W., et al., *Fisetin inhibits IL-1β-induced inflammatory response in human osteoarthritis chondrocytes through activating SIRT1 and attenuates the progression of osteoarthritis in mice.* Int Immunopharmacol, 2017. **45**: p. 135-147.

[96] Matsuzaki, T., et al., *Disruption of Sirt1 in chondrocytes causes accelerated progression of osteoarthritis under mechanical stress and during ageing in mice.* Ann Rheum Dis, 2014. **73**(7): p. 1397-404.

[97] Dvir-Ginzberg, M., et al., *Tumor necrosis factor α-mediated cleavage and inactivation of SirT1 in human osteoarthritic chondrocytes.* Arthritis Rheum, 2011. **63**(8): p. 2363-73.

[98] Batshon, G., et al., *Serum NT/CT SIRT1 ratio reflects early osteoarthritis and chondrosenescence.* Ann Rheum Dis, 2020. **79**(10): p. 1370-1380.

[99] Yang, H., et al., *Navitoclax (ABT263) reduces inflammation and promotes chondrogenic phenotype by clearing senescent osteoarthritic chondrocytes in osteoarthritis.* Aging (Albany NY), 2020. **12**(13): p. 12750-12770.

[100] Faust, H.J., et al., *IL-17 and immunologically induced senescence regulate response to injury in osteoarthritis.* J Clin Invest, 2020. **130**(10): p. 5493-5507.

[101] Farr, J.N., et al., *Identification of Senescent Cells in the Bone*

Microenvironment. J Bone Miner Res, 2016. **31**(11): p. 1920-1929.

[102] Piemontese, M., et al., *Old age causes de novo intracortical bone remodeling and porosity in mice.* JCI Insight, 2017. **2**(17).

[103] Farr, J.N. and S. Khosla, *Cellular senescence in bone.* Bone, 2019. **121**: p. 121-133.

[104] Farr, J.N., et al., *Targeting cellular senescence prevents age-related bone loss in mice.* Nat Med, 2017. **23**(9): p. 1072-1079.

[105] Lee, T.K., et al., *Optimal sites for forehead oscillometric blood pressure monitoring.* J Clin Monit, 1995. **11**(5): p. 298-304.

[106] Tu, K.N., et al., *Osteoporosis: A Review of Treatment Options.* P T, 2018. **43**(2): p. 92-104.

[107] McGreevy, C. and D. Williams, *Safety of drugs used in the treatment of osteoporosis.* Ther Adv Drug Saf, 2011. **2**(4): p. 159-72.

[108] Khosla, S., *Odanacatib: location and timing are everything.* J Bone Miner Res, 2012. **27**(3): p. 506-8.

[109] Coryell, P.R., B.O. Diekman, and R.F. Loeser, *Mechanisms and therapeutic implications of cellular senescence in osteoarthritis.* Nat Rev Rheumatol, 2021. **17**(1): p. 47-57.

[110] Cao, L., et al., *The use of autologous enriched bone marrow MSCs to enhance osteoporotic bone defect repair in long-term estrogen deficient goats.* Biomaterials, 2012. **33**(20): p. 5076-84.

[111] Freitag, J., et al., *Adipose derived mesenchymal stem cell therapy in the treatment of isolated knee chondral lesions: design of a randomised controlled pilot study comparing arthroscopic microfracture versus arthroscopic microfracture combined with postoperative mesenchymal stem cell injections.* BMJ Open, 2015. **5**(12): p. e009332.

[112] Matas, J., et al., *Umbilical Cord-Derived Mesenchymal Stromal Cells (MSCs) for Knee Osteoarthritis: Repeated MSC Dosing Is Superior to a Single MSC Dose and to Hyaluronic Acid in a Controlled Randomized Phase I/II Trial.* Stem Cells Transl Med, 2019. **8**(3): p. 215-224.

[113] Khaldoyanidi, S., *Directing stem cell homing.* Cell Stem Cell, 2008. **2**(3): p. 198-200.

[114] Sohni, A. and C.M. Verfaillie, *Mesenchymal stem cells migration homing and tracking.* Stem Cells Int, 2013. **2013**: p. 130763.

[115] Wynn, R.F., et al., *A small proportion of mesenchymal stem cells strongly expresses functionally active CXCR4 receptor capable of promoting migration to bone marrow.* Blood, 2004. **104**(9): p. 2643-5.

[116] Rüster, B., et al., *Mesenchymal stem cells display coordinated rolling and adhesion behavior on endothelial cells.* Blood, 2006. **108**(12): p. 3938-44.

[117] Docheva, D., et al., *Human mesenchymal stem cells in contact with their environment: surface characteristics and the integrin system.* J Cell Mol Med, 2007. **11**(1): p. 21-38.

[118] Teo, G.S., et al., *Mesenchymal stem cells transmigrate between and directly through tumor necrosis factor-α-activated endothelial cells via both leukocyte-like and novel mechanisms.* Stem Cells, 2012. **30**(11): p. 2472-86.

[119] Ullah, M., D.D. Liu, and A.S. Thakor, *Mesenchymal Stromal Cell Homing: Mechanisms and Strategies for Improvement.* iScience, 2019. **15**: p. 421-438.

[120] Spees, J.L., R.H. Lee, and C.A. Gregory, *Mechanisms of mesenchymal stem/stromal cell function.* Stem Cell Res Ther, 2016. **7**(1): p. 125.

[121] Mizuno, K., et al., *Exogenous synovial stem cells adhere to defect of meniscus and differentiate into cartilage cells.* J Med Dent Sci, 2008. **55**(1): p. 101-11.

[122] Park, Y.B., et al., *Single-stage cell-based cartilage repair in a rabbit model: cell tracking and in vivo chondrogenesis of human umbilical cord blood-derived mesenchymal stem cells and hyaluronic acid hydrogel composite.* Osteoarthritis Cartilage, 2017. **25**(4): p. 570-580.

[123] Mastrolia, I., et al., *Challenges in Clinical Development of Mesenchymal Stromal/Stem Cells: Concise Review.* Stem Cells Transl Med, 2019. **8**(11): p. 1135-1148.

[124] Pers, Y.M., et al., *Mesenchymal stem cells for the management of inflammation in osteoarthritis: state of the art and perspectives.* Osteoarthritis Cartilage, 2015. **23**(11): p. 2027-35.

[125] Izadpanah, R., et al., *Biologic properties of mesenchymal stem cells derived from bone marrow and adipose tissue.* J Cell Biochem, 2006. **99**(5): p. 1285-97.

[126] Satue, M., et al., *Intra-articularly injected mesenchymal stem cells promote cartilage regeneration, but do not permanently engraft in distant organs.* Sci Rep, 2019. **9**(1): p. 10153.

[127] Chiang, E.R., et al., *Allogeneic Mesenchymal Stem Cells in Combination with Hyaluronic Acid for the Treatment of Osteoarthritis in Rabbits.* PLoS One, 2016. **11**(2): p. e0149835.

[128] Song, F., et al., *Comparison of the efficacy of bone marrow mononuclear cells and bone mesenchymal stem cells in the treatment of osteoarthritis in a sheep model.* Int J Clin Exp Pathol, 2014. **7**(4): p. 1415-26.

[129] Ichioka, N., et al., *Prevention of senile osteoporosis in SAMP6 mice by intrabone marrow injection of allogeneic bone marrow cells.* Stem Cells, 2002. **20**(6): p. 542-51.

[130] Wang, Z., et al., *Efficacy of bone marrow-derived stem cells in strengthening osteoporotic bone in a rabbit model.* Tissue Eng, 2006. **12**(7): p. 1753-61.

[131] D'Ippolito, G., et al., *Age-related osteogenic potential of mesenchymal stromal stem cells from human vertebral bone marrow.* J Bone Miner Res, 1999. **14**(7): p. 1115-22.

[132] Stolzing, A., et al., *Age-related changes in human bone marrow-derived mesenchymal stem cells: consequences for cell therapies.* Mech Ageing Dev, 2008. **129**(3): p. 163-73.

[133] Zuk, P.A., et al., *Multilineage cells from human adipose tissue: implications for cell-based therapies.* Tissue Eng, 2001. **7**(2): p. 211-28.

[134] Kern, S., et al., *Comparative analysis of mesenchymal stem cells from bone marrow, umbilical cord blood, or adipose tissue.* Stem Cells, 2006. **24**(5): p. 1294-301.

[135] Koga, H., et al., *Local adherent technique for transplanting mesenchymal stem cells as a potential treatment of cartilage defect.* Arthritis Res Ther, 2008. **10**(4): p. R84.

[136] Kim, H.J. and G.I. Im, *Chondrogenic differentiation of adipose tissue-derived mesenchymal stem cells: greater doses of growth factor are necessary.* J Orthop Res, 2009. **27**(5): p. 612-9.

[137] Jurgens, W.J., et al., *Effect of tissue-harvesting site on yield of stem cells derived from adipose tissue: implications for cell-based therapies.* Cell Tissue Res, 2008. **332**(3): p. 415-26.

[138] Choudhery, M.S., et al., *Donor age negatively impacts adipose tissue-derived mesenchymal stem cell expansion and differentiation.* J Transl Med, 2014. **12**: p. 8.

[139] Tang, Y., et al., *A comparative assessment of adipose-derived stem cells*

from subcutaneous and visceral fat as a potential cell source for knee osteoarthritis treatment. J Cell Mol Med, 2017. **21**(9): p. 2153-2162.

[140] Kuroda, K., et al., *The paracrine effect of adipose-derived stem cells inhibits osteoarthritis progression.* BMC Musculoskelet Disord, 2015. **16**: p. 236.

[141] Mirsaidi, A., et al., *Therapeutic potential of adipose-derived stromal cells in age-related osteoporosis.* Biomaterials, 2014. **35**(26): p. 7326-35.

[142] Cho, S.W., et al., *Human adipose tissue-derived stromal cell therapy prevents bone loss in ovariectomized nude mouse.* Tissue Eng Part A, 2012. **18**(9-10): p. 1067-78.

[143] Ye, X., et al., *Adipose-derived stem cells alleviate osteoporosis by enhancing osteogenesis and inhibiting adipogenesis in a rabbit model.* Cytotherapy, 2014. **16**(12): p. 1643-55.

[144] De Bari, C., et al., *Multipotent mesenchymal stem cells from adult human synovial membrane.* Arthritis Rheum, 2001. **44**(8): p. 1928-42.

[145] Li, N., et al., *Synovial membrane mesenchymal stem cells: past life, current situation, and application in bone and joint diseases.* Stem Cell Res Ther, 2020. **11**(1): p. 381.

[146] Nakamura, T., et al., *Arthroscopic, histological and MRI analyses of cartilage repair after a minimally invasive method of transplantation of allogeneic synovial mesenchymal stromal cells into cartilage defects in pigs.* Cytotherapy, 2012. **14**(3): p. 327-38.

[147] Sasaki, A., et al., *Canine mesenchymal stem cells from synovium have a higher chondrogenic potential than those from infrapatellar fat pad, adipose tissue, and bone marrow.* PLoS One, 2018. **13**(8): p. e0202922.

[148] To, K., et al., *Synovium-Derived Mesenchymal Stem Cell Transplantation in Cartilage Regeneration: A PRISMA Review of in vivo Studies.* Front Bioeng Biotechnol, 2019. **7**: p. 314.

[149] Ozeki, N., et al., *Not single but periodic injections of synovial mesenchymal stem cells maintain viable cells in knees and inhibit osteoarthritis progression in rats.* Osteoarthritis Cartilage, 2016. **24**(6): p. 1061-70.

[150] Sakaguchi, Y., et al., *Comparison of human stem cells derived from various mesenchymal tissues: superiority of synoium as a cell source.* Arthritis Rheum, 2005. **52**(8): p. 2521-9.

[151] Qu-Petersen, Z., et al., *Identification of a novel population of muscle stem cells in mice: potential for muscle regeneration.* J Cell Biol, 2002. **157**(5): p. 851-64.

[152] Gharaibeh, B., et al., *Isolation of a slowly adhering cell fraction containing stem cells from murine skeletal muscle by the preplate technique.* Nat Protoc, 2008. **3**(9): p. 1501-9.

[153] Lavasani, M., et al., *Isolation of muscle-derived stem/progenitor cells based on adhesion characteristics to collagen-coated surfaces.* Methods Mol Biol, 2013. **976**: p. 53-65.

[154] Camernik, K., et al., *Skeletal-muscle-derived mesenchymal stem/stromal cells from patients with osteoarthritis show superior biological properties compared to bone-derived cells.* Stem Cell Res, 2019. **38**: p. 101465.

[155] Peng, H., et al., *Synergistic enhancement of bone formation and healing by stem cell-expressed VEGF and bone morphogenetic protein-4.* J Clin Invest, 2002. **110**(6): p. 751-9.

[156] Payne, T.R., et al., *A relationship between vascular endothelial growth factor, angiogenesis, and cardiac repair after muscle stem cell transplantation into*

ischemic hearts. J Am Coll Cardiol, 2007. **50**(17): p. 1677-84.

[157] Oshima, H., et al., *Differential myocardial infarct repair with muscle stem cells compared to myoblasts.* Mol Ther, 2005. **12**(6): p. 1130-41.

[158] Deasy, B.M., et al., *A role for cell sex in stem cell-mediated skeletal muscle regeneration: female cells have higher muscle regeneration efficiency.* J Cell Biol, 2007. **177**(1): p. 73-86.

[159] Lee, J.Y., et al., *Clonal isolation of muscle-derived cells capable of enhancing muscle regeneration and bone healing.* J Cell Biol, 2000. **150**(5): p. 1085-100.

[160] Matsumoto, T., et al., *The influence of sex on the chondrogenic potential of muscle-derived stem cells: implications for cartilage regeneration and repair.* Arthritis Rheum, 2008. **58**(12): p. 3809-19.

[161] Scibetta, A.C., et al., *Characterization of the chondrogenic and osteogenic potential of male and female human muscle-derived stem cells: Implication for stem cell therapy.* J Orthop Res, 2019. **37**(6): p. 1339-1349.

[162] Corsi, K.A., et al., *Osteogenic potential of postnatal skeletal muscle-derived stem cells is influenced by donor sex.* J Bone Miner Res, 2007. **22**(10): p. 1592-602.

[163] Meszaros, L.B., et al., *Effect of host sex and sex hormones on muscle-derived stem cell-mediated bone formation and defect healing.* Tissue Eng Part A, 2012. **18**(17-18): p. 1751-9.

[164] Kuroda, R., et al., *Cartilage repair using bone morphogenetic protein 4 and muscle-derived stem cells.* Arthritis Rheum, 2006. **54**(2): p. 433-42.

[165] Matsumoto, T., et al., *Cartilage repair in a rat model of osteoarthritis through intraarticular transplantation of muscle-derived stem cells expressing bone morphogenetic protein 4 and soluble Flt-1.* Arthritis Rheum, 2009. **60**(5): p. 1390-405.

[166] Gao, X., et al., *Influences of donor and host age on human muscle-derived stem cell-mediated bone regeneration.* Stem Cell Res Ther, 2018. **9**(1): p. 316.

[167] Koh, Y.G. and Y.J. Choi, *Infrapatellar fat pad-derived mesenchymal stem cell therapy for knee osteoarthritis.* Knee, 2012. **19**(6): p. 902-7.

[168] Vega, A., et al., *Treatment of Knee Osteoarthritis With Allogeneic Bone Marrow Mesenchymal Stem Cells: A Randomized Controlled Trial.* Transplantation, 2015. **99**(8): p. 1681-90.

[169] Pers, Y.M., et al., *Adipose Mesenchymal Stromal Cell-Based Therapy for Severe Osteoarthritis of the Knee: A Phase I Dose-Escalation Trial.* Stem Cells Transl Med, 2016. **5**(7): p. 847-56.

[170] Jo, C.H., et al., *Intra-articular Injection of Mesenchymal Stem Cells for the Treatment of Osteoarthritis of the Knee: A 2-Year Follow-up Study.* Am J Sports Med, 2017. **45**(12): p. 2774-2783.

[171] Bastos, R., et al., *Intra-articular injections of expanded mesenchymal stem cells with and without addition of platelet-rich plasma are safe and effective for knee osteoarthritis.* Knee Surg Sports Traumatol Arthrosc, 2018. **26**(11): p. 3342-3350.

[172] Emadedin, M., et al., *Intra-articular implantation of autologous bone marrow-derived mesenchymal stromal cells to treat knee osteoarthritis: a randomized, triple-blind, placebo-controlled phase 1/2 clinical trial.* Cytotherapy, 2018. **20**(10): p. 1238-1246.

[173] Khalifeh Soltani, S., et al., *Safety and efficacy of allogenic placental mesenchymal stem cells for treating knee osteoarthritis: a pilot study.* Cytotherapy, 2019. **21**(1): p. 54-63.

[174] Jones, I.A., et al., *A randomized, controlled study to evaluate the efficacy of*

intra-articular, autologous adipose tissue injections for the treatment of mild-to-moderate knee osteoarthritis compared to hyaluronic acid: a study protocol. BMC Musculoskelet Disord, 2018. **19**(1): p. 383.

[175] Gupta, P.K., et al., *Efficacy and safety of adult human bone marrow-derived, cultured, pooled, allogeneic mesenchymal stromal cells (Stempeucel®): preclinical and clinical trial in osteoarthritis of the knee joint.* Arthritis Res Ther, 2016. **18**(1): p. 301.

[176] Kuah, D., et al., *Safety, tolerability and efficacy of intra-articular Progenza in knee osteoarthritis: a randomized double-blind placebo-controlled single ascending dose study.* J Transl Med, 2018. **16**(1): p. 49.

[177] Song, Y., et al., *Human adipose-derived mesenchymal stem cells for osteoarthritis: a pilot study with long-term follow-up and repeated injections.* Regen Med, 2018. **13**(3): p. 295-307.

[178] Spasovski, D., et al., *Intra-articular injection of autologous adipose-derived mesenchymal stem cells in the treatment of knee osteoarthritis.* J Gene Med, 2018. **20**(1).

[179] Wyles, C.C., et al., *Mesenchymal stem cell therapy for osteoarthritis: current perspectives.* Stem Cells Cloning, 2015. **8**: p. 117-24.

[180] Sekiya, I., et al., *Arthroscopic Transplantation of Synovial Stem Cells Improves Clinical Outcomes in Knees With Cartilage Defects.* Clin Orthop Relat Res, 2015. **473**(7): p. 2316-26.

[181] Shimomura, K., et al., *First-in-Human Pilot Study of Implantation of a Scaffold-Free Tissue-Engineered Construct Generated From Autologous Synovial Mesenchymal Stem Cells for Repair of Knee Chondral Lesions.* Am J Sports Med, 2018. **46**(10): p. 2384-2393.

Chapter 4

Strategies to Treat Pulmonary Hypertension Using Programmed Cell Death-Inducing Anti-Cancer Drugs without Damaging the Heart

Yuichiro J. Suzuki, Yasmine F. Ibrahim, Vladyslava Rybka, Jaquantey R. Bowens, Adenike S. Falade and Nataliia V. Shults

Abstract

Pulmonary arterial hypertension (PAH) is a fatal disease without a cure. By the time patients are diagnosed with PAH, thickening of pulmonary arterial (PA) walls and the narrowing of vascular lumen have already developed due to the abnormal growth of pulmonary vascular cells, contributing to the elevated pulmonary vascular resistance and the right ventricle (RV) damage. Therefore, agents that eliminate excess pulmonary vascular wall cells have therapeutic potential, and the apoptosis-based therapy using anti-cancer drugs may be promising for the treatment of PAH. However, cell death agents could also exert adverse effects including cardiotoxicity, complicating the development of such therapies for PAH patients who already have the damaged heart. We tested the concept that programmed cell death-inducing anti-cancer drugs may reduce the PA wall thickening using rat models of PAH. We found that: (i) The treatment of PAH animals with anthracycline-, proteasome inhibitor- or Bcl-2 inhibitor-classes of anti-cancer drugs after the pulmonary vascular remodeling had already developed resulted in the reversal of PA wall thickening and opened up the lumen; (ii) These effects were accompanied by the apoptosis of PA wall cells in PAH rats, but not in normal healthy rats, suggesting the anti-cancer drugs selectively kill remodeled vascular cells; (iii) The RV affected by PAH was not further damaged by anthracyclines or proteasome inhibitors; (iv) While the left ventricle (LV) was damaged by these drugs, we identified cardioprotective agents that protect the heart against drug-induced cell death without affecting the efficacy to reverse the PA remodeling; and (v) docetaxel, not only reversed pulmonary vascular remodeling without exerting RV or LV toxicity, but also repaired the RV damage caused by PAH. Thus, the inclusion of programmed cell death-inducing anti-cancer drugs should be considered for treating PAH patients.

Keywords: anti-cancer drugs, apoptosis, autophagy, heart, programmed cell death, pulmonary hypertension, vascular remodeling

1. Introduction

Pulmonary arterial hypertension (PAH) is a fatal disease that can affect both females and males of any age including children. If untreated, increased pulmonary vascular resistance results in right heart failure and kills patients within several years [1, 2]. Even with the currently available therapeutic drugs that are mainly vasodilators, the survival duration of the patients remains unacceptably short [3, 4]. It has been reported that the median survival for patients diagnosed with PAH is 2.8 years from the time of diagnosis (3-year survival: 48%) if untreated [5, 6]. Even with currently available therapies, only 58–75% of PAH patients survive for 3 years [7–10]. PAH is a progressive disease, and by the time patients are diagnosed, thickening of pulmonary artery (PA) walls and the narrowing of vascular lumen have already developed due to the abnormal growth of pulmonary vascular cells, contributing to the elevated pulmonary vascular resistance and the right ventricle (RV) damage. Therefore, agents that eliminate excess pulmonary vascular wall cells have therapeutic potential, and we hypothesize that the programmed cell death-based therapy using anti-cancer drugs would help treat PAH patients [11]. However, cell death agents could also exert adverse effects including cardiotoxicity, complicating the development of such therapies for PAH patients with the already damaged heart.

2. Anti-cancer drugs reverse pulmonary vascular remodeling

In our earlier study, we found that an anthracycline anti-cancer drug daunorubicin (DNR) is an effective agent that can cause apoptosis of cultured PA smooth muscle cells (PASMCs) [11, 12]. Based on these results, we hypothesized that the administration of DNR to rats would result in the reversal of pulmonary vascular remodeling. In these experiments, Sprague-Dawley (SD) rats were treated with chronic hypoxia (10% oxygen) for 2 weeks to promote the thickening of PA medial walls. After the PA wall thickening was developed, rats were injected with DNR and maintained in the hypoxia condition for 3 days. As shown in hematoxylin and eosin (H&E) stain images of **Figure 1A**, DNR effectively reduced the PA wall thickness [13]. Similarly, in this study, another class of anti-cancer drugs, proteasome inhibitors such as MG132 and bortezomib (**Figure 1B**) also reduced the PA wall thickening in the chronic hypoxia model of pulmonary hypertension (PH) in rats [13].

An animal model, in which SD rats are injected with SU5416 and exposed to hypoxia promoting severe PAH with pulmonary vascular lesions resembling those of humans [14], has become a gold standard to study PAH [15]. The experimental design often involves a single subcutaneous injection of SU5416, followed by subjecting the animals to chronic hypoxia for 3 weeks. Subsequently, the animals are kept in normoxia, and severe PAH and pulmonary vascular remodeling are progressively developed. We found that programmed cell death-inducing anti-cancer drugs reversed pulmonary vascular remodeling in this model of PAH as well. **Figure 1C** shows the results of our experiments, in which another proteasome inhibitor, carfilzomib (CFZ) injected 4 times over two weeks after the pulmonary vascular remodeling was developed effectively reduced the PA wall thickness in PAH rats [16]. Proteasome inhibition-dependent reversal of pulmonary vascular remodeling occurred through the reduction of both intimal and medial wall thickening, suggesting that both endothelial cells and smooth muscle cells (SMCs) can be affected by these anti-cancer drugs [13].

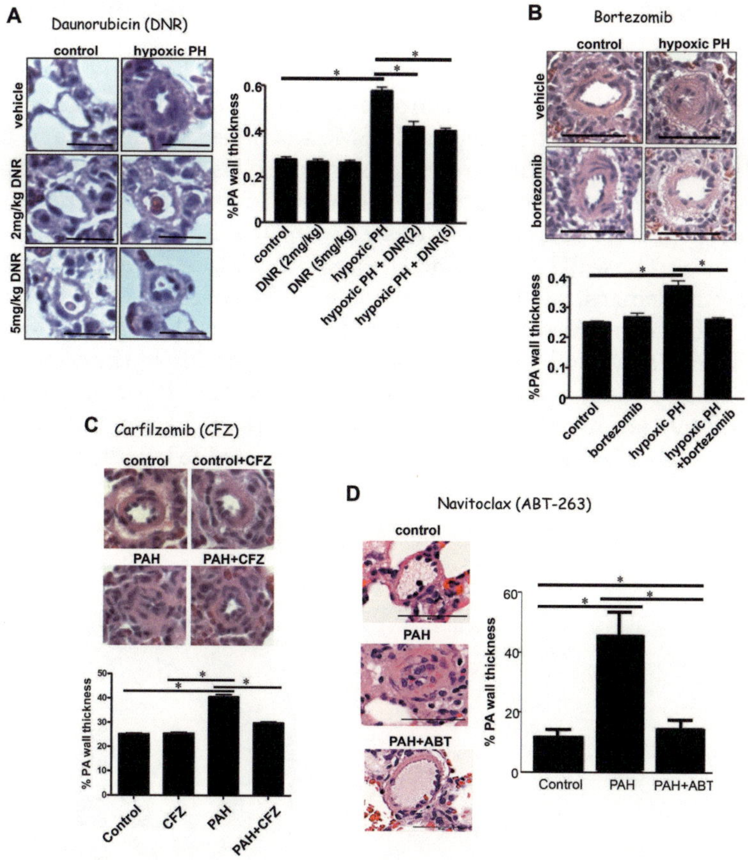

Figure 1.
*Effects of programmed cell death-inducing anti-cancer drugs on pulmonary vascular remodeling. (A & B) SD rats were treated with chronic hypoxia for 2 weeks to produce pulmonary vascular thickening and injected with DNR or bortezomib. Rats were then placed back in the hypoxic environment. Three days after the injection, lungs were harvested and H&E staining was performed (Adapted from Ibrahim et al. [13] with permission). (C & D) SD rats were subjected to SU5416/hypoxia to promote PAH. After pulmonary vascular remodeling was developed, rats were injected with CFZ or navitoclax twice a week for 2 weeks. Lungs were harvested and H&E staining was performed (Adapted from Wang et al. [16] and Rybka et al. [17] with permission). Bar graphs represent means ± SEM of % PA wall thickness. * denotes that the values are significantly different from each other at $P < 0.05$.*

While anthracyclines such as DNR and proteasome inhibitors are effective inducers of apoptosis of PASMCs [11, 13, 16], these agents could exert other biologic actions. Thus, we tested the effects of a more 'pure' apoptosis inducer, navitoclax (ABT-263) that inhibits anti-apoptotic proteins Bcl-2 and Bcl-x_L. We found that this drug also reversed PA remodeling in SD rats as well as in Fischer rats with PAH promoted by SU5416 + hypoxia (**Figure 1D**; [17]). The reversal of pulmonary vascular remodeling by navitoclax was also recently reported by van der Feen et al. [18] in a different experimental model of PAH in rats.

These results provided important information, in live experimental animals, showing that programmed cell death-inducing anti-cancer drugs are capable of reversing pulmonary vascular remodeling in multiple models of PH. While this knowledge established a basis for exploring whether causing the death of pulmonary vascular cells clinically benefits PAH patients, it also generated many questions that need to be addressed.

3. Susceptibility of normal and diseased cells toward apoptosis-inducing anti-cancer drugs

One question is whether both the proliferative synthetic phenotype and the differentiated contractile phenotype of PASMCs are killed by these drugs. It is preferable that only abnormally grown cells are killed, as it is important to preserve contractile SMCs that are needed for the pulmonary circulatory system to function.

The examination of PAs from rats treated with DNR by terminal deoxynucleotidyl transferase dUTP nick end labeling (TUNEL) staining, which detects apoptotic cells, demonstrated that only remodeled PAs of rats with PH exhibited apoptotic cells, but not healthy control rats (**Figure 2A**; [13]). Similar results were obtained in the analysis of cleaved caspase-3 as an indication of the occurrence of apoptotic cells by Western blotting. As shown in **Figure 2B**, PAs from rats treated with chronic hypoxia to promote PH and subsequently treated with DNR exhibited significantly higher levels of cleaved caspase-3 compared to healthy rats injected

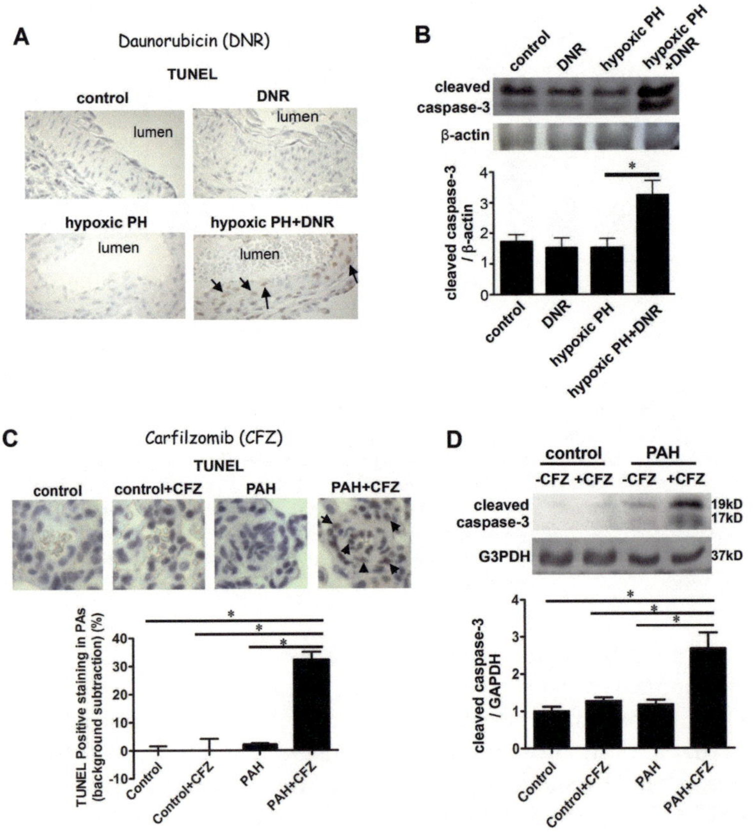

Figure 2.

*Effects of programmed cell death-inducing anti-cancer drugs on apoptosis. (A & B) SD rats were treated with chronic hypoxia for 2 weeks to produce pulmonary vascular thickening and injected with DNR. Rats were then placed back in the hypoxic environment. Three days after the injection, lungs were harvested and TUNEL staining and Western blotting with the cleaved caspase-3 antibody were performed to monitor apoptosis (Adapted from Ibrahim et al. [13] with permission). (C & D) SD rats were subjected to SU5416/hypoxia to promote PAH. After pulmonary vascular remodeling was developed, rats were injected with CFZ twice a week for 2 weeks. Lungs were harvested and TUNEL staining and Western blotting with the cleaved caspase-3 antibody were performed to monitor apoptosis (Adapted from Wang et al. [16] with permission). Bar graphs represent means ± SEM. * denotes that the values are significantly different from each other at P < 0.05.*

Figure 3.
Effects of genestein on proliferating and differentiated human PASMC apoptosis. Proliferating human PASMCs purchased from Cell Applications grown in Human SMC Growth Medium and differentiated PASMCs produced using the Human SMC Differentiation Medium were treated with genistein (50 μM). Apoptotic cells were identified by green fluorescence produced using the DePsipher Mitochondrial Potential assay.

with DNR [13]. These results revealed that unwanted abnormally grown pulmonary vascular cells can preferentially be killed by this anti-cancer drug.

Results shown in **Figure 2C** and **D** demonstrated that this increased susceptibility of pulmonary vascular cells in PH animals can also be seen with another anti-cancer drug. CFZ also caused the apoptosis in PAs of rats with PAH induced by SU5416/hypoxia, while no apoptosis signals were observed in control healthy rats treated with CFZ as monitored by TUNEL assay (**Figure 2C**) and Western blotting using the cleaved caspase-3 antibody (**Figure 2D**) [16].

We hypothesized that anti-cancer drugs preferentially kill the proliferating phenotype of SMCs over differentiated SMCs. Our experiments using cultured PASMCs showed that only proliferating SMCs, but not differentiated SMCs, were killed by DNR [13]. **Figure 3** shows similar experimental results when proliferating and differentiated human PASMCs were treated with genistein, a naturally occurring isoflavone. DePsipher Mitochondrial Potential assay (Trevigen, Gaithersburg, MD, USA) showed that green fluorescent apoptotic cells were only observed when proliferating PASMCs were treated with genistein, while differentiated PASMCs produced by using the Differentiation Medium (Cell Applications, Inc., San Diego, CA, USA) were resistant to be killed by the same concentration of genistein. These results demonstrate that proliferating PASMCs are more susceptible to undergo apoptosis compared to differentiated PASMCs, suggesting that apoptosis-inducing drugs eliminated unwanted proliferating PASMCs while preserving the contractile phenotypic cells with muscle functions.

4. Role of autophagic cell death

One interesting observation we came across in relation to the mechanism of PASMC killing by anthracycline- and proteasome inhibitor-classes of anti-cancer drug is that, in addition, to apoptosis, another programmed cell death mechanism, namely autophagic cell death is also involved. We initially found that autophagy of the cells is increased in PAs of PH rats treated with DNR [13]. Similar results were observed in cultured proliferating human PASMCs when cells were treated with DNR. Further, DNR-induced cell killing was attenuated when an autophagy mediator, LC3B, was knocked down [13]. CFZ-induced cell killing was also found to involve autophagy, and we further identified the role of tumor protein p53-inducible nuclear protein 1 (TP53INP1) in this mechanism [16].

5. The ability of the remodeled RV to cope with program cell death-inducing drugs

Drugs that promote programmed cell death are effective anti-cancer drugs, however, they also exert serious potentially life-threatening complications [19]. Cardiotoxicity is a major complication that accompanies the use of anti-cancer drugs especially anthracyclines. Since PAH patients already have the weakened heart, the use of these anti-cancer drugs would be considered to be contraindications. However, we found that the RV affected by PAH is remarkably resistant to drug-induced myocardial cell killing. As we characterized the RV of PAH rats injected with DNR to reverse PA remodeling as described above, we found that DNR administration to PAH rats did not influence the RV contractility or the RV structure [13]. This study also found that DNR did not promote apoptosis of cardiomyocytes in hypertrophied RV in rats with PH (**Figure 4A**; [13]). Similarly, CFZ that was found to effectively reverse PA remodeling did not cause apoptosis in the RV in SU5416/hypoxia model of PAH in rats (**Figure 4B**; [16]). These are highly significant findings revealing that the RV affected by PAH is resistant to DNR and CFZ, drugs that are known induce cardiotoxicity and cardiomyocyte killing in the normal heart, providing evidence that the clinical use of these anti-cancer drugs in PAH patients may not be contraindications.

By contrast, bortezomib was found to promote apoptosis in both RV and left ventricle (LV) of rats with PH induced by monocrotaline [20]. Also, navitoclax (ABT-263; an inhibitor of anti-apoptotic proteins, Bcl-2 and Bcl-xL), not only

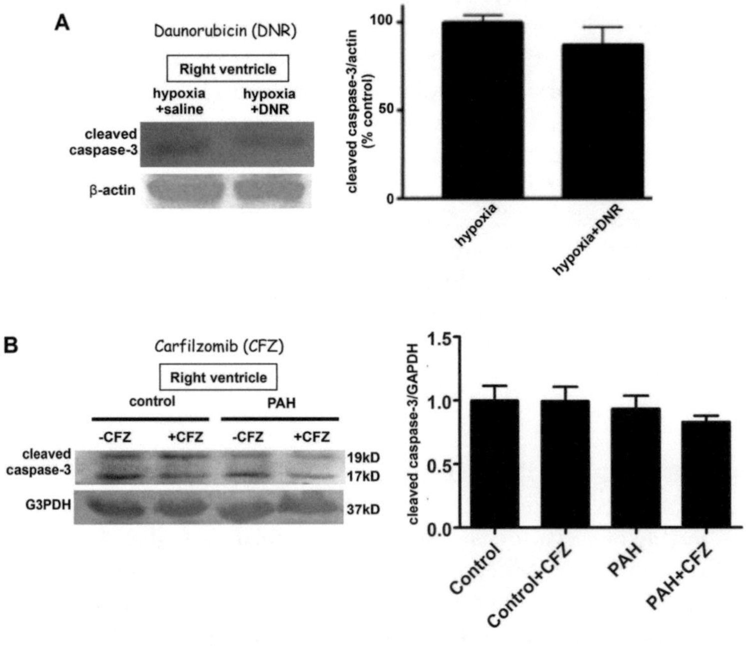

Figure 4.

Effects of programmed cell death-inducing anti-cancer drugs on the RV affected by PAH. (A) SD rats were treated with chronic hypoxia for 2 weeks to produce pulmonary vascular thickening and injected with DNR. Rats were then placed back in the hypoxic environment. Three days after the injection, RV tissues were harvested and Western blotting with the cleaved caspase-3 antibody was performed to monitor apoptosis (Adapted from Ibrahim et al. [13] with permission). (B) SD rats were subjected to SU5416/hypoxia to promote PAH. After pulmonary vascular remodeling was developed, rats were injected with CFZ twice a week for 2 weeks. RV tissues were harvested and Western blotting with the cleaved caspase-3 antibody were performed to monitor apoptosis (Adapted from Wang et al. [16] with permission). Bar graphs represent means ± SEM. All the values were not significantly different from each other at $P < 0.05$.

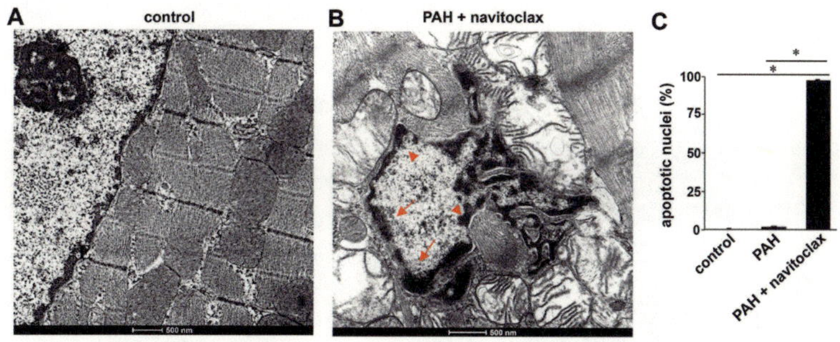

Figure 5.
*Navitoclax promotes cardiomyocyte apoptosis in the RV affected by PAH. (A) The normal RV structure of RV cardiomyocytes as monitored by transmission electron microscopy. Magnification ×16,000. (B) SD rats were subjected to SU5416/hypoxia to promote PAH. After pulmonary vascular remodeling was developed, rats were injected with navitoclax twice a week for 2 weeks as described in Rybka et al. [17]. RV tissues were harvested and a transmission electron microscope was used to analyze the apoptotic nuclei. Bar graphs represent means ± SEM. * denotes that the values are significantly different from each other at P < 0.05. Transmission electron microscopy studies were performed as described in Shults et al. [32].*

caused apoptosis in the remodeled PA [17], but also promoted apoptosis in RV myocytes in PAH rats. **Figure 5** shows the transmission electron microscopy images of normal SD rat RV myocytes (**Figure 5A**) and RV myocytes from PAH SD rats treated with navitoclax exhibiting signs of apoptosis (**Figure 5B**). The nuclei PAH rats treated with navitoclax underwent the fragmentation with dramatic changes in the nuclear chromatin with the segregated heterochromatin that distributed preferentially within the nuclear envelope as sharply defined clumped bodies (**Figure 5B**, red arrowheads). The quantification of apoptotic nuclei revealed that the most of RV myocytes in PAH rats became apoptotic when treated with navitoclax (**Figure 5C**).

These results suggest that, while three classes of anti-cancer drugs have so far been found to be effective in reversing PA remodeling, hypertrophied RV myocytes are only resistant to DNR and CFZ, while Bcl-2/Bcl-x_L inhibition seems target downstream of apoptotic pathway thus escapes from the resistance to cardiomyocyte killing. Whether the RV damaging effects of bortezomib in PAH rats [20] is specific to the model induced by monocrotaline that can exert non-specific pathophysiologic actions need further investigations, however, the data so far do not support the use of bortezomib in the PAH treatment. CFZ that is considered to be a safe alternative to bortezomib in cancer therapy [21] and is a more selective and irreversible inhibitor of the chymotrypsin-like activity of the 20S proteasome [22, 23] could be more promising.

6. Cardioprotective agents to cope with LV myocyte death by anti-cancer drugs

Our laboratory previously found a cell-signaling pathway for the downregulation of Bcl-x_L/Bcl-2 that results in the apoptosis of cardiomyocytes [24]. This pathway was found as a consequence of our laboratory cloning the promoter region of the GATA4 transcription factor that regulates gene transcription of Bcl-x_L and Bcl-2. We found that CBF/NF-Y binding to the CCAAT box of the Gata4 promoter is inhibited by DNR through the activation of p53 in cardiomyocytes [24], but not in PASMCs [13]. Thus, we hypothesized that p53 inhibitors would protect the heart against cardiotoxicity induced by anti-cancer drugs without affecting the efficacy of these drugs to reverse PA remodeling.

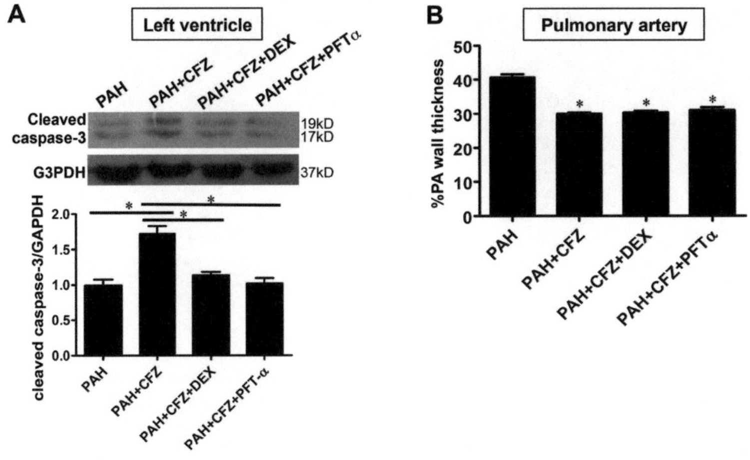

Figure 6.
*Dexrazoxane (DEX) and pifithrin-α (PFT-α) protects the LV from CFZ-induced apoptosis without affecting the efficacy of CFZ in reversing PA remodeling. (A) PAH rats (SU5416/hypoxia) were divided into 4 groups. DEX or PFT-α was injected intraperitoneally along with CFZ, twice a week for two weeks. Rats were then sacrificed 3 days after the last injection. LV tissues were homogenized, and subjected to Western blotting for the cleaved capase-3 formation. The bar graph represents means ± SEM. * denotes that the values are significantly different from each other at P < 0.05. (B) The reduction of remodeled PA thickness induced by CFZ was not affected by DEX or PFT-α by analyzing H&E staining. The bar graph represents means ± SEM. * denotes that the values are significantly different from the PAH value at P < 0.05. (Adapted from Wang et al. [16] with permission).*

In our study of CFZ as described above, we found that this proteasome inhibitor is effective in reversing PA remodeling and that the RV affected by PAH is resistant to CFZ toxicity [16]. However, as expected from the earlier cancer studies, CFZ did cause the cardiomyocyte apoptosis in the LV of PAH rats (**Figure 6A**). As a support for our hypothesis, this CFZ-induced apoptosis of LV cardiomyocytes was inhibited by a p53 inhibitor, pifithrin-α in PAH rats (**Figure 6A**), while this cardioprotective agent did not interfere with CFZ reducing the PA wall thickening (**Figure 6B**). Interestingly, we found that a clinically used cardioprotective drug, dexrazoxane, also protected that LV of PAH rats from CFZ toxicity without affecting the reversal of PA remodeling. Further investigations are needed to determine whether these actions of dexrazoxane involve p53. Nevertheless, these results suggest including dexrazoxane or a p53 inhibitor to protect the LV against drug-induced damage while treating PAH patients with anti-cancer drugs.

7. Docetaxel as a fascinating drug that reduces pulmonary vascular wall thickening and repairs the damaged right ventricle

Since experiments described above provided results that support the use of anti-cancer drugs to reverse pulmonary vascular remodeling, we further searched for other drugs that could be useful. In an effort to find effective drugs that preferentially kill proliferating PASMCs, we screened various drugs [25]. We found that docetaxel (a taxane class of anti-cancer drugs that stabilizes and inhibits microtubules) effectively killed proliferating human PASMCs, but not differentiated human PASMCs in culture [25]. As we tested docetaxel for reversing pulmonary vascular remodeling in the SU5416/hypoxia model of PAH, we found that this drug indeed was effective in reducing thickened pulmonary vascular walls (**Figure 7A**). Effects were similar to anthracycline-, proteasome inhibitor-, and Bcl-2/Bcl-x_L

inhibitor-classes of drugs. As described above, we found that DNR and CFZ did not have adverse effects on the hypertrophied RV in PAH rats while Bcl-2/Bcl-x_L inhibition resulted in the apoptosis of RV myocytes. Docetaxel also did not exhibit adverse effects on the hypertrophied RV in PAH rats. Moreover, this drug repaired damaged RV caused by PAH. In SU5416/hypoxia model of PAH, the RV was found to have significant cardiac fibrosis as shown in the blue stain of Masson's trichrome staining in **Figure 7B**. Remarkably, these fibrotic lesions were eliminated by the treatment of PAH rats with docetaxel (**Figure 7B**; [25]).

These results suggest that docetaxel is an effective drug that can reverse pulmonary vascular remodeling and at the same time it can also repair the damaged RV caused by PAH at least in SD rats treated with SU5416/hypoxia. Another taxane drug, paclitaxel has also been shown to attenuate pulmonary vascular remodeling in rodent models of PAH induced by monocrotaline or SU5416/hypoxia [26–30]. However, the ability of paclitaxel to repair the RV in PAH animals has not been reported. It is interesting to note that paclitaxel has been shown to improve cardiac function during ischemia in isolated rat and rabbit hearts [31], reinforcing the idea that taxanes have the capacity to promote cardiac repair.

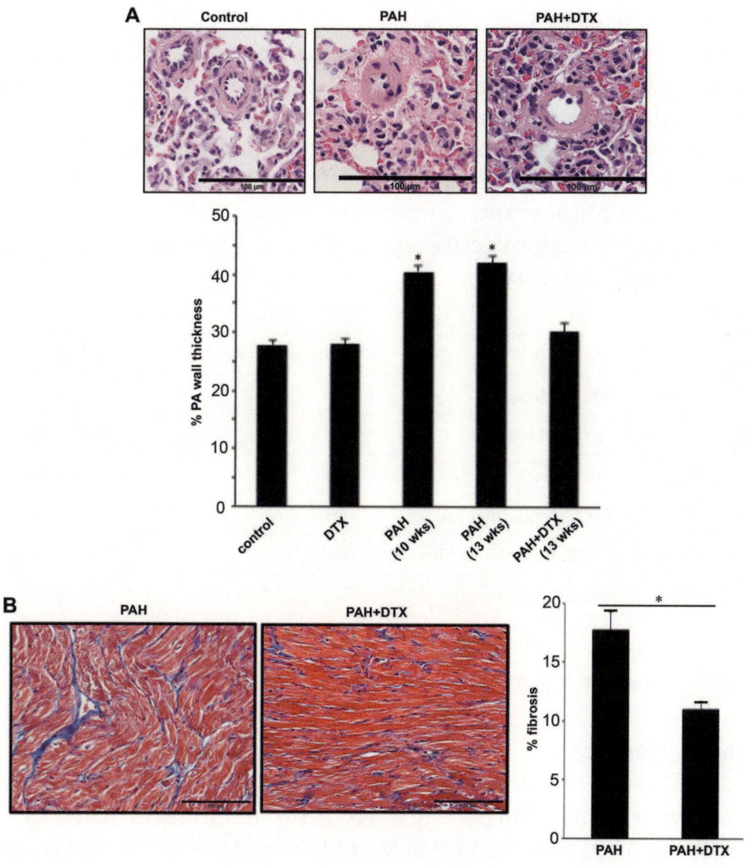

Figure 7.
*Docetaxel reverses pulmonary vascular remodeling and cardiac fibrosis in the RV in PAH rats. SD rats were subjected to SU5416/hypoxia to promote PAH. After pulmonary vascular remodeling was developed, rats were injected with DTX twice a week for 2 weeks. (A) Lungs were harvested and H&E staining was performed. The bar graph represents means ± SEM of % PA wall thickness. * denotes that the values are significantly different from each other at P < 0.05. (B) Heart tissues were harvested and Masson's trichrome staining was performed to monitor fibrosis. The bar graph represents means ± SEM of % fiborsis in the RV. * denotes that the values are significantly different from the PAH value at P < 0.05. (Adapted from Ibrahim et al. [13] with permission).*

8. Conclusions

We tested the concept that cell death-inducing anti-cancer drugs may reduce the PA wall thickening using rat models of PAH. We found that: (1) The treatment of PAH rats with anthracycline-, proteasome inhibitor- or Bcl-2/Bcl-x_L inhibitor-classes of drugs after pulmonary vascular remodeling had occurred resulted in the reversal of pulmonary vascular remodeling and opened up the lumen; (2) These effects were accompanied by the apoptosis of PA wall cells in PAH rats, but these drugs did not promote apoptosis in normal healthy rats, suggesting the anti-cancer drugs selectively kill remodeled vascular cells; (3) DNR, an anthracycline, and CFZ, a proteasome inhibitor, did not adversely affect the hypertrophied RV of PAH rats. (4) While the LV was damaged by CFZ, we identified cardioprotective agents (dexrazoxane and pifithrin-alpha) that can protect the heart against drug-induced cell death without affecting the efficacy of the drugs to reduce PA remodeling; (5) Docetaxel, a taxane class of anti-cancer drugs, not only reversed pulmonary vascular remodeling without exerting RV or LV toxicity, but also repaired the RV damaged caused by PAH. These findings from our laboratory as well as reports by other laboratories on the topic of the effects of programmed cell death-inducing anti-cancer drugs on remodeled PA and the RV affected by PAH in experimental animals are summarized in **Table 1**.

These results demonstrate that certain anti-cancer drugs effectively and selectively cause programmed cell death of abnormally grown cells in the remodeled pulmonary vasculature without adversely affecting the RV in rat models of PAH. Thus, the inclusion of programmed cell death-induced anti-cancer drugs may be promising for treating PAH patients. Human clinical trials of PAH treatment that test the effectiveness of these anti-cancer drugs as mono-therapies or combination therapies along with cardioprotective agents described here as well as already available vasodilators are warranted.

	Reduces remodeled PA	Affects remodeled RV
Daunorubicin, DNR (Anthracycline)	Yes	No effects
Carfilzomib, CFZ (Proteasome inhibitor)	Yes	No effects
Bortezomib (Proteasome inhibitor)	Yes	Apoptosis
Navitoclax, ABT-263 (Bcl-2/Bcl-x_L inhibitor)	Yes	Apoptosis
Docetaxel, DTX (Taxane; Microtubule inhibitor)	Yes	Repairs
Paclitaxel (Taxane; Microtubule inhibitor)	Yes	Unknown

Table 1.
Abilities of various anti-cancer drugs to affect PA and RV remodeling.

Acknowledgements

This work was supported in part by the NIH (grant numbers R01HL072844, R21AI142649, R03AG059554, and R03AA026516) to Y.J.S. The content is solely the responsibility of the authors and does not necessarily represent the official views of the NIH.

Conflict of interest

None.

Author details

Yuichiro J. Suzuki[1]*, Yasmine F. Ibrahim[2], Vladyslava Rybka[1], Jaquantey R. Bowens[1], Adenike S. Falade[1] and Nataliia V. Shults[1]

1 Department of Pharmacology and Physiology, Georgetown University Medical Center, Washington, DC, USA

2 Department of Pharmacology, Minia University Faculty of Medicine, Minia, Egypt

*Address all correspondence to: ys82@georgetown.edu

References

[1] Delcroix M, Naeije R. Optimising the management of pulmonary arterial hypertension patients: emergency treatments. Eur Respir Rev. 2010;19:204-211.

[2] McLaughlin VV, Shah SJ, Souza R, Humbert M. Management of pulmonary arterial hypertension. J Am Coll Cardiol. 2015;65:1976-1997.

[3] Fallah F. Recent strategies in treatment of pulmonary arterial hypertension, a review. Glob J Health Sci. 2015;7:307-322.

[4] Rosenkranz S. Pulmonary hypertension 2015: current definitions, terminology, and novel treatment options. Clin Res Cardiol. 2015;104:197-207.

[5] D'Alonzo GE, Barst RJ, Ayres SM, Bergofsky EH, Brundage BH, Detre KM, Fishman AP, Goldring RM, Groves BM, Kernis JT, et al. Survival in patients with primary pulmonary hypertension. Results from a national prospective registry. Ann Intern Med. 1991;115:343-349.

[6] Runo JR, Loyd JE. Primary pulmonary hypertension. Lancet. 2003;361:1533-1544.

[7] Benza RL, Miller DP, Frost A, Barst RJ, Krichman AM, and McGoon MD. Analysis of the lung allocation score estimation of risk of death in patients with pulmonary arterial hypertension using data from the REVEAL Registry. Transplantation. 2010;90:298-305.

[8] Humbert M, Sitbon O, Yaïci A, Montani D, O'Callaghan DS, Jaïs X, Parent F, Savale L, Natali D, Günther S, Chaouat A, Chabot F, Cordier JF, Habib G, Gressin V, Jing ZC, Souza R, Simonneau G; French Pulmonary Arterial Hypertension Network. Survival in incident and prevalent cohorts of patients with pulmonary arterial hypertension. Eur Respir J. 2010;36:549-555.

[9] Thenappan T, Shah SJ, Rich S, Tian L, Archer SL, Gomberg-Maitland M. Survival in pulmonary arterial hypertension: a reappraisal of the NIH risk stratification equation. Eur Respir J. 2010;35:1079-1087.

[10] Olsson KM, Delcroix M, Ghofrani HA, Tiede H, Huscher D, Speich R, Grünig E, Staehler G, Rosenkranz S, Halank M, Held M, Lange TJ, Behr J, Klose H, Claussen M, Ewert R, Opitz CF, Vizza CD, Scelsi L, Vonk-Noordegraaf A, Kaemmerer H, Gibbs JS, Coghlan G, Pepke-Zaba J, Schulz U, Gorenflo M, Pittrow D, Hoeper MM. Anticoagulation and survival in pulmonary arterial hypertension: results from the Comparative, Prospective Registry of Newly Initiated Therapies for Pulmonary Hypertension (COMPERA). Circulation. 2014;129:57-65.

[11] Suzuki YJ, Nagase H, Wong CM, Kumar SV, Jain V, Park AM, Day RM. Regulation of Bcl-x_L expression in lung vascular smooth muscle. Am J Respir Cell Mol Biol. 2007;36:678-687.

[12] Suzuki YJ, Ibrahim YF, Shults NV. Apoptosis-based therapy to treat pulmonary arterial hypertension. J Rare Dis Res Treat. 2016;1:17-24.

[13] Ibrahim YF, Wong CM, Pavlickova L, Liu L, Trasar L, Bansal G, Suzuki YJ. Mechanism of the susceptibility of remodeled pulmonary vessels to drug-induced cell killing. J Am Heart Assoc. 2014;3:e000520.

[14] Taraseviciene-Stewart L, Kasahara Y, Alger L, Hirth P, Mc

Mahon G, Waltenberger J, Voelkel NF, Tuder RM. Inhibition of the VEGF receptor 2 combined with chronic hypoxia causes cell death-dependent pulmonary endothelial cell proliferation and severe pulmonary hypertension. FASEB J. 2001;15:427-438.

[15] Oka M, Homma N, Taraseviciene-Stewart L, Morris KG, Kraskauskas D, Burns N, Voelkel NF, McMurtry IF. Rho kinase-mediated vasoconstriction is important in severe occlusive pulmonary arterial hypertension in rats. Circ Res. 2007;100:923-929.

[16] Wang X, Ibrahim YF, Das D, Zungu-Edmondson M, Shults NV, Suzuki YJ. Carfilzomib reverses pulmonary arterial hypertension. Cardiovasc Res. 2016;110:188-199.

[17] Rybka V, Suzuki YJ, Shults NV. Effects of Bcl-2/Bcl-x_L inhibitors on pulmonary artery smooth muscle cells. Antioxidants. 2018;7:150.

[18] van der Feen DE, Bossers GPL, Hagdorn QAJ, Moonen JR, Kurakula K, Szulcek R, Chappell J, Vallania F, Donato M, Kok K, Kohli JS, Petersen AH, van Leusden T, Demaria M, Goumans MTH, De Boer RA, Khatri P, Rabinovitch M, Berger RMF, Bartelds B. Cellular senescence impairs the reversibility of pulmonary arterial hypertension. Sci Transl Med. 2020;12:eaaw4974.

[19] Vincent DT, Ibrahim YF, Espey MG, Suzuki YJ. The role of antioxidants in the era of cardio-oncology. Cancer Chemother Pharmacol. 2013;72:1157-1168.

[20] Kim SY, Lee JH, Huh JW, Kim HJ, Park MK, Ro JY, Oh YM, Lee SD, Lee YS. Bortezomib alleviates experimental pulmonary arterial hypertension. Am J Respir Cell Mol Biol. 2012;47:698-708.

[21] Herndon TM, Deisseroth A, Kaminskas E, Kane RC, Koti KM, Rothmann MD, Habtemariam B, Bullock J, Bray JD, Hawes J, Palmby TR, Jee J, Adams W, Mahayni H, Brown J, Dorantes A, Sridhara R, Farrell AT, Pazdur R. U.S. Food and Drug Administration approval: carfilzomib for the treatment of multiple myeloma. Clin Cancer Res. 2013;19:4559-4563.

[22] Demo SD, Kirk CJ, Aujay MA, Buchholz TJ, Dajee M, Ho MN, Jiang J, Laidig GJ, Lewis ER, Parlati F, Shenk KD, Smyth MS, Sun CM, Vallone MK, Woo TM, Molineaux CJ, Bennett MK. Antitumor activity of PR-171, a novel irreversible inhibitor of the proteasome. Cancer Res. 2007;67:6383-6391.

[23] Kortuem KM, Stewart AK. Carfilzomib. Blood. 2013;121:893-897.

[24] Park AM, Nagase H, Liu L, Vinod Kumar S, Szwergold N, Wong CM, Suzuki YJ. Mechanism of anthracycline-mediated down-regulation of GATA4 in the heart. Cardiovasc Res. 2011;90:97-104.

[25] Ibrahim YF, Shults NV, Rybka V, Suzuki YJ. Docetaxel reverses pulmonary vascular remodeling by decreasing autophagy and resolves right ventricular fibrosis. J Pharmacol Exp Ther. 2017;363:20-34.

[26] Yin Y, Wu X, Yang Z, Zhao J, Wang X, Zhang Q, Yuan M, Xie L, Liu H, He Q. The potential efficacy of R8-modified paclitaxel-loaded liposomes on pulmonary arterial hypertension. Pharm Res. 2013;30:2050-2062.

[27] Savai R, Al-Tamari HM, Sedding D, Kojonazarov B, Muecke C, Teske R, Capecchi MR, Weissmann N, Grimminger F, Seeger W, Schermuly RT, Pullamsetti SS. Pro-proliferative and inflammatory signaling converge

on FoxO1 transcription factor in pulmonary hypertension. Nat Med. 2014;20:1289-1300.

[28] Feng W, Wang J, Yan X, Zhai C, Shi W, Wang Q, Zhang Q, Li M. Paclitaxel alleviates monocrotaline-induced pulmonary arterial hypertension via inhibition of FoxO1-mediated autophagy. Naunyn Schmiedebergs Arch Pharmacol. 2019;392:605-613.

[29] Zhao J, Yang M, Wu X, Yang Z, Jia P, Sun Y, Li G, Xie L, Liu B, Liu H. Effects of paclitaxel intervention on pulmonary vascular remodeling in rats with pulmonary hypertension. Exp Ther Med. 2019;17:1163-1170.

[30] Kassa B, Mickael C, Kumar R, Sanders L, Koyanagi D, Hernandez-Saavedra D, Tuder RM, Graham BB. Paclitaxel blocks Th2-mediated TGF-β activation in Schistosoma mansoni-induced pulmonary hypertension. Pulm Circ. 2019;9:2045894018820813.

[31] Xiao J, Zhao H, Liang D, Liu Y, Zhang H, Liu Y, Li J, Peng L, Zhou Z, Chen YH. Taxol, a microtubule stabilizer, improves cardiac contractile function during ischemia in vitro. Pharmacology. 2010;85:301-310.

[32] Shults NV, Kanovka SS, Ten Eyck JE, Rybka V, Suzuki YJ. Ultrastructural changes of the right ventricular myocytes in pulmonary arterial hypertension. J Am Heart Assoc. 2019;8:e011227.

Section 2

Hypothenar Muscles

Chapter 5

Hypothenar Muscles and Guyon's Canal

Georgi P. Georgiev

Abstract

The increased number of articles in the last years about hypothenar variations and some misdescriptions and the role of the additional structures to ulnar nerve and artery compression, as well as my unostentatious contribution in the field, provoked me to write this chapter. The aim of it is to present in detail the usual hypothenar muscular anatomy, including the origins and insertions of the hypothenar muscles, their relations to each other, the vascular supply and innervation, the function of the muscles, the reported variations and their possible clinical implications. Herein, I also presented briefly the Guyon's canal anatomy and some interesting comments about it. Presenting the compendium about hypothenar muscles and the canal to my opinion will help the anatomists and the clinicians to better understand the clinically oriented anatomy. They also will be more qualified in the anatomical dissection course as well as during the surgical interventions. The detailed knowledge of the anatomy in the region would be also useful to medical students in better understanding the hypothenar region.

Keywords: hypothenar muscles, anatomy, variations, Guyon's canal, clinical significance

1. Introduction

Muscular variations of the hypothenar have been well described in the medical literature [1–3]. In most cases, these muscles are asymptomatic and are found during anatomical dissections, surgical interventions or imaging modalities. In some cases, variant hypothenar muscles may cause ulnar nerve and artery compression, as presented in some surgical case reports [4–10]. Shea and McClain [11] reported that around 3% of compression neuropathies are due to variant muscle.

At the region of the wrist, the ulnar nerve passes through a fibro-osseous tunnel known as "Guyon's canal" or "distal ulnar tunnel", in which the ulnar nerve could be compressed [4, 6, 8, 12–14]. Different causes as trauma, lipoma, false aneurysm of the ulnar artery, ganglion cyst and rarely aberrant muscular slips have been reported [3, 15]. In some surgical reports, different variant muscles, usually abductor digiti minimi, followed by flexor digiti minimi brevis have been reported, as a cause of ulnar nerve compression [4, 6, 8, 12–14]. In all cases the excision of the additional muscle was curative. However, it should be pointed out that the clinical appearance of the variant muscles should be related to two factors: the anatomical site and the muscle hypertrophy [4]. According to Turner and Caird [16], a provoking factor, such as acute injury or repetitive minor trauma, as well as the type of the work, could predispose to hypertrophy of the variant muscles. There are reports of ulnar nerve compression syndromes provoked by anomalous muscles in which additional factors exist [1, 4, 17]. In addition to nerve compression, a hypothenar muscle variation could be also associated with thrombosis of the ulnar artery in Guyon's canal [7, 18].

In recent years the increased interest and numerous articles about hypothenar variants and Guyon's canal provoke me to make a brief review and make a compendium of normal anatomy of the muscles in the hypothenar region as well to present the reported variations and their possible clinical significance. I hope that this chapter will make future studies on this theme easier and help anatomist, hand surgeons and medical students for better knowledge of anatomy and for better clinical practice.

2. Anatomy, variations and clinical significance

2.1 Palmaris brevis muscle (Pbm)

Pbm is a small cutaneous hand muscle, located superficially to the hypothenar eminence and considered to be atavistic remnant of the panniculus carnosus. The Pbm is usually presented by a thin, quadrilateral in form muscle body, lying beneath the skin of the ulnar aspect of the palm. It has been reported to start from the flexor retinaculum and palmar aponeurosis and insert in the skin and fascia of the medial palm (**Figure 1**) [19–24].

2.1.1 Variations

The variations of the Pbm are rarely reported in the literature. This muscle could vary in size and may be absent or duplicated; it could also insert to flexor digiti minimi brevis muscle or the pisiform bone [25, 26].

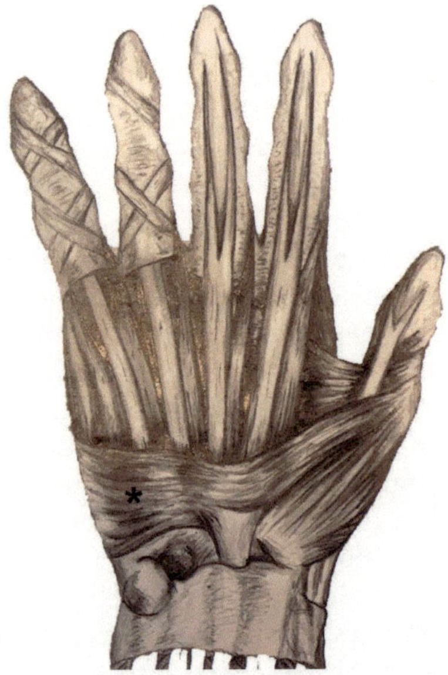

Figure 1.
Schematic anterior view of the wrist and hand presenting the Pbm (asterisk).

2.1.2 Clinical application

This muscle and its variants could simulate soft tissue tumor [27] or provoke ulnar nerve compression at the wrist [17, 28]; Pbm flap could be used in the treatment of recurrent carpal tunnel syndrome [29] and Pbm spasm syndrome [30].

2.1.3 Actions

Pbm deepens the hollow of the palm and presents muscular barrier of the ulnar neurovascular bundle at the wrist [21].

2.1.4 Innervation

The Pbm is innervated by the motor component of the superficial branch of the ulnar nerve [24].

2.2 Abductor digiti minimi muscle (ADMm)

The ADMm is situated more ulnarly of the hypothenar muscles and arises from the pisiform bone and from the tendon of the flexor carpi ulnaris muscle; it attaches as a flat tendon that finally divides into two parts: one that inserts to the ulnar side of the base of the proximal phalanx of the little finger and the other to the ulnar border of the aponeurosis of the extensor digiti quinti proprius (**Figure 2**) [20, 21].

2.2.1 Variations

Different variations of ADMm have been reported in the anatomical and surgical literature. They include the absence and presence of a second head, variant origin (from the pisiform bone, fascia of the forearm, palmaris longus tendon,

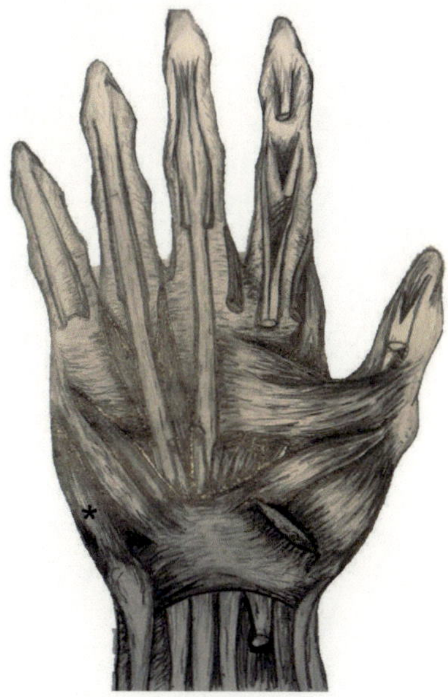

Figure 2.
Schematic anterior view of the wrist and hand presenting the ADMm (asterisk).

fascia of the flexor carpi radialis, intermuscular fascia, flexor carpi ulnaris, flexor retinaculum, both from the flexor retinaculum and antebrachial fascia), fusion with the flexor digiti minimi brevis, presence of a "deep abductor-flexor"(m. abductor-flexor digiti minimi profundus), triple origin and also coexistence with reversed palmaris longus muscle [4, 5, 9, 15, 25–27, 31–38].

2.2.2 Clinical application

Hypertrophied ADMm could simulate soft tissue tumor [5] or ulnar nerve compression at the wrist [13, 14] and may be associated with ulnar artery thrombosis in Guyon's canal [7]; ADMm myocutaneous flap can be used for opponensplasty [39].

2.2.3 Actions

The ADMm abducts the little finger from the ring finger and contributes to the fifth metacarpophalangeal joint flexion and interphalangeal extension [20, 21].

2.2.4 Innervation

ADMm is innervated by the deep branch of the ulnar nerve (C8, Th1) [20, 21].

2.3 Flexor digiti minimi brevis muscle (FDMBm)

The FDMBm is situated more radially than the ADMm. It starts from the hamulus of the hamate bone, and the anterior surface of the flexor retinaculum, and inserts into the ulnar side of the base of the phalanx of the little finger (**Figure 3**) [20, 21].

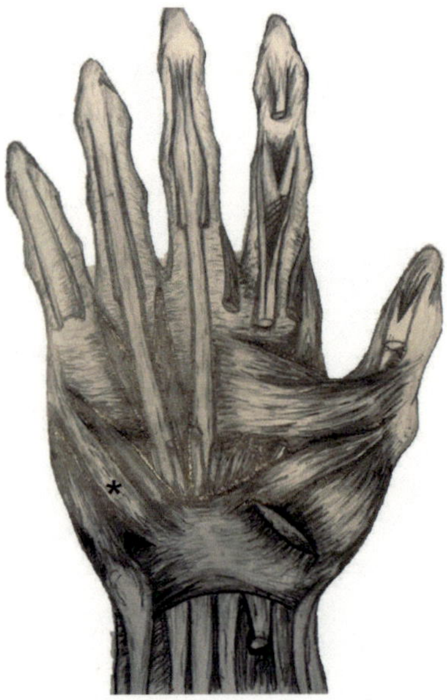

Figure 3.
Schematic anterior view of the wrist and hand presenting the FDMBm (asterisk).

2.3.1 Variations

The reported variations of the FDMBm in the available literature are absence and presence of an accessory slip from the palmaris longus tendon, presence of a slip to the metacarpal, replacement by a tendinous band arising from the flexor carpi ulnaris muscle and inserting into the fifth proximal phalanx and the hamate bone, presence of accessory FDMBm, fusion with ADMm or opponens digiti minimi muscles, origin from the antebrachial fascia, deep abductor-flexor of little finger, FDMBm with triple origin and origin from flexor carpi radialis muscle [25, 26, 33, 40–43].

2.3.2 Clinical application

Hypertrophied FDBMm could simulate soft tissue tumor [44] or ulnar nerve compression at the wrist [8, 12].

2.3.3 Actions

The FDBMm flexion of the proximal phalanx, also with some lateral rotation [20, 21].

2.3.4 Innervation

FDBMm is innervated by the deep branch of the ulnar nerve (C8, Th1) [20, 21].

2.4 Opponens digiti minimi muscle (ODMm)

The ODMm has a triangular form, lying beneath the ADMm and FDMBm. It starts from the hamulus of the hamate bone and near part of the flexor retinaculum and attaches to the ulnar margin and palmar surface of the fifth metacarpal bone (**Figure 4**) [20, 21].

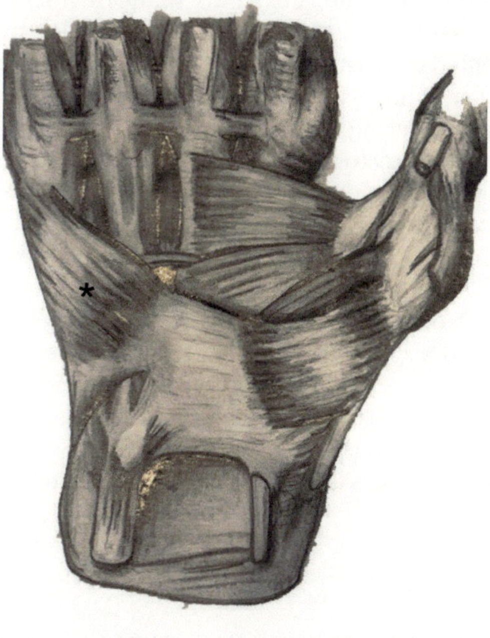

Figure 4.
Schematic anterior view of the wrist and hand presenting the ODMm (asterisk).

2.4.1 Variations

The reported variations of the ODMm are rarely described in the literature and include absence, splitting into two parts and merging with ADMm [25, 26].

2.4.2 Clinical application

There is no clinical application reported.

2.4.3 Innervation

The ODMm is innervated by the deep branch of the ulnar nerve (C8, Th1) [20, 21].

2.4.4 Actions

The ODMm flexes and laterally rotates the fifth metacarpal bone at the carpometacarpal joint, brings the fifth finger into opposition with the thumb and together with the ADMm and FDMBm absorbs forces on the ulnar border of the hand [20, 21].

3. Anatomy of Guyon's canal

The Guyon's canal or ulnar canal is a fibro-osseous tunnel situated between the pisiform and the hook of the hamate and measured around 40–45 mm in length [45, 46]. In 1861, it is first described by the French surgeon Jean Casimir Félix Guyon. He presented it as an intra-aponeurotic compartment which the anterior wall is constituted by a fibrous layer and its posterior wall by the anterior carpal ligament [47]. Guyon's canal is situated between the proximal edge of the palmar carpal ligament to the fibrous arch of the hypothenar muscles at the level of the hook of the hamate (**Figure 5**). Through this canal the ulnar nerve and artery pass from the forearm to the palm, as the nerve is lying deep and ulnar to the artery. Of course, the vena comitans and connective fatty tissue fill up this space [48, 49].

Guyon's canal has been described as a space with complex and variable anatomy [48]. It should be pointed out that the canal is not a rigid structure and varying in its dimensions [50]. According to Ombaba et al. [49], it has dynamic space, and its relationships are changed during wrist movement.

Precise knowledge of the anatomy of Guyon's canal is essential in understanding the diagnosis and treatment of ulnar tunnel syndrome [51]. This tunnel is a potential entrapment site that could provoke compression of the ulnar nerve, presented by paresthesia or numbness, or both, to the ulnar two digits, as well as hand weakness, atrophy and ungainliness [51]. The compression could be localized in three zones [11, 52]:

Zone I: compression occurs proximal to or within Guyon's canal, before the nerve bifurcation, and presents with combined motor and sensory deficits.

Zone II: compression involves only the deep motor branch and occurs as the ulnar nerve exits Guyon's canal or at the hook of the hamate level.

Zone III: compression, with isolated superficial sensory branch involvement, may also occur as the nerve exits Guyon's canal or at the hook of the hamate distal to the bifurcation.

Importantly during surgical interventions, all three compartments should be decompressed, including the pisohamate hiatus, by releasing the pisohamate arcade [49].

Different muscular variations have been reported as the most common anatomical predispositions that might contribute to the ulnar nerve compression

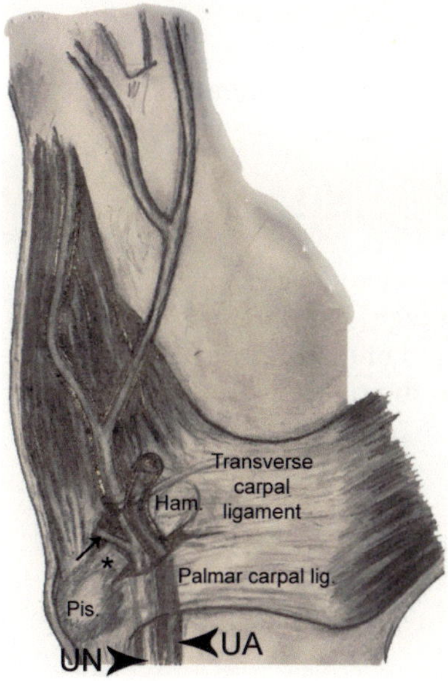

Figure 5.
Anatomy of Guyon's canal. Pisohamate ligament (asterisk); deep motor branch of the ulnar nerve (arrow); pisiform bone (Pis.); hamate bone (ham.); ulnar nerve (UN); ulnar artery (UA).

in Guyon's canal [51]. Different muscular variations related to the ADMm, followed by FDMBm, are reported [8, 12–14]. In these cases, ultrasonography or MRI could help clinicians to clearly identify the muscular variants in Guyon's canal [38, 53].

4. Conclusion

This chapter summarizes the existing data in the literature concerning the anatomy of the hypothenar muscles, as well as its variants and the anatomy of Guyon's canal. I hope that the presented literature data will help students in learning anatomy, help the anatomists and hand surgeons during their works, as well as for better scientific production and to better understand, describe and classify the muscular variations and Guyon's canal. Using the chapter, they will avoid mistakes in classifying different variations of the hypothenar muscles. I think that the knowledge of reported variants will ensure self-confidence and avoid confusions during wrist and palm procedures, especially during releasing Guyon's canal, extended carpal tunnel release, or palmar fasciectomy. Precise knowledge of the anatomy and its variant also would help radiologist in their routine work on ultrasonography and MRI, where different variations could be evaluated in cases of soft tissue tumor or in diagnosis of primary or recurrent ulnar nerve compression.

Acknowledgements

The author would like to thank Mr. Ahmed Hashim Mohamed Ali (Student at Medical University of Sofia, Bulgaria) for the kind help through his excellent drawings illustrated in the chapter.

Conflict of interest

The author declares no conflict of interest.

Abbreviations

Pbm	palmaris brevis muscle
ADMm	abductor digiti minimi muscle
FDMBm	flexor digiti minimi brevis muscle
ODMm	opponens digiti minimi muscle

Author details

Georgi P. Georgiev
Department of Orthopedics and Traumatology, University Hospital Queen Giovanna-ISUL, Medical University of Sofia, Sofia, Bulgaria

*Address all correspondence to: georgievgp@yahoo.com

References

[1] Al-Qattan MM. Ulnar nerve compression at the wrist by the accessory abductor digiti minimi muscle: Wrist trauma as a precipitating factor. Hand Surgery. 2004;**9**:79-82. DOI: 10.1142/s0218810404001899

[2] Murata K, Tamai M, Gupta A. Anatomic study of variations of hypothenar muscles and arborization patterns of the ulnar nerve in the hand. The Journal of Hand Surgery. 2004;**29**:500-509. DOI: 10.1016/j.jhsa.2004.01.006

[3] Claassen H, Schmitt O, Schulze M, Wree A. Variation in the hypothenar muscles and its impact on ulnar tunnel syndrome. Surgical and Radiologic Anatomy. 2013;**35**:893-899. DOI: 10.1007/s00276-013-1113-5

[4] Jeffery AK. Compression of the deep palmar branch of the ulnar nerve by an anomalous muscle. Case report and review. Journal of Bone and Joint Surgery. British Volume (London). 1971;**53**:718-723

[5] Simodynes EE, Cochran RM 2nd. Anomalous muscles in the hand and wrist--report of three cases. The Journal of Hand Surgery. 1981;**6**(6):553-554. DOI: 10.1016/s0363-5023(81)80129-3

[6] James MR, Rowley DI, Norris SH. Ulnar nerve compression by an accessory abductor digiti minimi muscle presenting following injury. Injury. 1987;**18**:66-67. DOI: 10.1016/0020-1383(87)90393-7

[7] Pribyl CR, Moneim MS. Anomalous hand muscle found in the Guyon's canal at exploration for ulnar artery thrombosis. A case report. Clinical Orthopaedics and Related Research. 1994;**306**:120-123

[8] Spinner RJ, Lins RE, Spinner M. Compression of the medial half of the deep branch of the ulnar nerve by an anomalous origin of the flexor digiti minimi. A case report. The Journal of Bone and Joint Surgery. American Volume. 1996;**78**:427-430

[9] Slavchev SA, Georgiev GP. Aberrant abductor digiti minimi muscle found during open surgical decompression of the carpal tunnel: Case report. Revista Argentina de Anatomía Clínica. 2013;**5**:88-91

[10] Georgiev GP. Anatomical variations of muscles in the human body and their relevance for clinical practice. International Journal of Anatomical Variations. 2018;**11**:48-49

[11] Shea JD, McClain EJ. Ulnar-nerve compression syndromes at and below the wrist. The Journal of Bone and Joint Surgery. American Volume. 1969;**51**:1095-1103

[12] Sälgeback S. Ulnar tunnel syndrome caused by anomalous muscles. Case report. Scandinavian Journal of Plastic and Reconstructive Surgery. 1977;**11**:255-258

[13] Coraci D, Luchetti R, Paolasso I, Santilli V, Padua L. Intermittent ulnar nerve compression due to accessory abductor digiti minimi muscle: Crucial diagnostic role of nerve ultrasound. Muscle & Nerve. 2015;**52**:463-464. DOI: 10.1002/mus.24660

[14] Mohan J, Ramesh BA. Ulnar nerve compression by accessory abductor digiti minimi muscle. Nigerian Journal of Plastic Surgery. 2016;**12**:69-71. DOI: 10.4103/0794-9316.202440

[15] Georgiev GP, Jelev L. Unusual coexistence of a variant abductor digiti minimi and reversed palmaris longus and their possible relation to median and ulnar nerves entrapment at the wrist. Romanian Journal of Morphology and Embryology. 2009;**50**(4):725-727

[16] Turner MS, Caird DM. Anomalous muscles and ulnar nerve compression at the wrist. The Hand. 1977;**9**:140-142. DOI: 10.1016/s0072-968x(77)80007-7

[17] Tonkin MA, Lister GD. The palmaris brevis profundus. An anomalous muscle associated with ulnar nerve compression at the wrist. The Journal of Hand Surgery. 1985;**10**:862-864

[18] Moss DP, Forthman CL. Ulnar artery thrombosis associated with anomalous hypothenar muscle. Journal of Surgical Orthopaedic Advances. 2008 Summer;**17**(2):85-88

[19] von Lanz T, Wachsmuth W. Arm. In: Praktische Anatomie. 1. Bd, 3. Teil. Berlin: Verlag von Julius Springer; 1935. p. 193

[20] Clemente CD. Anatomy of the Human Body. 30th ed. Philadelphia: Lea and Febiger; 1985. pp. 553-554

[21] Standring S, Borley NR, Gray H. Gray's Anatomy: The Anatomical Basis of Clinical Practice. 40th ed. Edinburgh: Churchill Livingstone/Elsevier; 2008. pp. 857-898

[22] Nayak SR, Krishnamurthy A. An unusually large palmaris brevis muscle and its clinical significance. Clinical Anatomy. 2007;**20**:978-979. DOI: 10.1002/ca.20542

[23] Moore CW, Beveridge TS, Rice CL. Fiber type composition of the palmaris brevis muscle: Implications for palmar function. Journal of Anatomy. 2017;**231**:626-633. DOI: 10.1111/joa.12652

[24] Moore CW, Rice CL. Structural and functional anatomy of the palmaris brevis: Grasping for answers. Journal of Anatomy. 2017;**231**:939-946. DOI: 10.1111/joa.12675

[25] Macalister A. Additional observations on muscular anomalies in human anatomy (third series), with a catalogue of the principal muscular variations hitherto published. Transactions of the Royal Irish Academy. 1875;**25**:1-130

[26] Bergman RA, Afifi AK, Miyauchi R. Part I: Muscular system. In: Illustrated Encyclopedia Of Human Anatomic Variations. 2019. Available from: http://www.anatomyatlases.org/AnatomicVariants/AnatomyHP.shtml [Accessed: 26 November 2019]

[27] Lipscomb PR. Duplication of hypothenar muscles simulating soft-tissue tumor of the hand. Report of a case. The Journal of Bone and Joint Surgery. American Volume. 1960;**42**:1058-1061

[28] Robinson D, Aghasi MK, Halperin N. Ulnar tunnel syndrome caused by an accessory palmaris muscle. Orthopaedic Review. 1989;**18**:345-347

[29] Rose EH. The use of the palmaris brevis flap in recurrent carpal tunnel syndrome. Hand Clinics. 1996;**12**: 389-395

[30] Eswaradass PV, Kalidoss R, Ramasamy B, Gnanashanmugham G. Familial palmaris brevis spasm syndrome. Annals of Indian Academy of Neurology. 2014;**17**:141-142. DOI: 10.4103/0972-2327.128597

[31] Wood J. Variations in human myology observed during the winter session of 1866-1867 at King's college, London. In Variations in Human Myology. Vol IV. Proceedings of the Royal Society of London. 1867;**15**:518-546

[32] Wood J. Variations in human myology observed during the winter season of 1867-1868 at King's college, London. In Variations in Human Myology. Vol XVII. Proceedings of the Royal Society of London. 1868;**16**:483-525

[33] Le Double A. Muscles de la main. In: Traité des variations du système musculaire de l'homme. Tome II. Paris: Schleicher Frères; 1897. pp. 153-218

[34] Gloobe H, Pecket P. An anomalous muscle in the canal of Guyon (a possible ulnar nerve compression). Anatomischer Anzeiger. 1973;**133**: 477-479

[35] Sañudo JR, Mirapeix RM, Ferreira B. A rare anomaly of abductor digiti minimi. Journal of Anatomy. 1993;**182**(3):439-442

[36] Georgiev GP, Jelev L, Surchev L. Undescribed variant muscle - "deep abductor-flexor" of the little finger, in relation to ulnar nerve compression at the wrist. Annals of Anatomy. 2007;**189**:276-282. DOI: 10.1016/j.aanat.2006.11.003

[37] Georgiev GP, Jelev L, Surchev L. Variations of the hypothenar muscles. Acta Morphologica et Anthropologica. 2008;**13**:313-315

[38] Georgiev GP, Jelev L, Kinov P. Aberrant muscles at the Guyon's canal. International Journal of Anatomical Variations. 2010;**4**:67-69

[39] Upton J, Taghinia AH. Abductor digiti minimi myocutaneous flap for opponensplasty in congenital hypoplastic thumbs. Plastic and Reconstructive Surgery. 2008;**122**:1807-1811. DOI: 10.1097/PRS.0b013e31818cc260

[40] Saadeh FA, Bergman RA. An unusual accessory flexor (opponens) digiti minimi muscle. Anatomischer Anzeiger. 1988;**165**:327-329

[41] Wingerter S, Gupta S, Le S, Shamasunder S, Bernstein R, Rabitaille W, et al. Unusual origin of the flexor digiti minimi brevis muscle. Clinical Anatomy. 2003;**16**:531-533. DOI: 10.1002/ca.10122

[42] Georgiev GP, Jelev L. Variant triple origin of the flexor digiti minimi brevis (Manus) muscle in relation to ulnar nerve and artery compression at the wrist. Clinical Anatomy. 2007;**20**: 976-977. DOI: 10.1002/ca.20529

[43] Georgiev GP, Jelev L. An aberrant flexor digiti minimi brevis Manus muscle. The Journal of Hand Surgery. 2011;**36**:1965-1967. DOI: 10.1016/j.jhsa.2011.09.002

[44] Harvie P, Patel N, Ostlere SJ. Prevalence and epidemiological variation of anomalous muscles at Guyon's canal. Journal of Hand Surgery (British). 2004;**29**:26-29. DOI: 10.1016/j.jhsb.2003.08.004

[45] Depukat P, Mizia E, Kuniewicz M, Bonczar T, Mazur M, Pełka P, et al. Syndrome of canal of Guyon - definition, diagnosis, treatment and complication. Folia Medica Cracoviensia. 2015;**55**:17-23

[46] Depukat P, Henry BM, Popieluszko P, Roy J, Mizia E, Konopka T, et al. Anatomical variability and histological structure of the ulnar nerve in the Guyon's canal. Archives of Orthopaedic and Trauma Surgery. 2017;**137**:277-283. DOI: 10.1007/s00402-016-2616-4

[47] Guyon F. Note sur une disposition anatomique propere a la face anterieure de la region du poignet et non encore decrite. Bulletins de la Société Anatomique de Paris. 1861;**6**:184-186

[48] Bachoura A, Jacoby SM. Ulnar tunnel syndrome. The Orthopedic Clinics of North America. 2012;**43**: 467-474. DOI: 10.1016/j.ocl.2012.07.016

[49] Ombaba J, Kuo M, Rayan G. Anatomy of the ulnar tunnel and the influence of wrist motion on its morphology. The Journal of Hand Surgery. 2010;**35**:760-768. DOI: 10.1016/j.jhsa.2010.02.028

[50] Fadel ZT, Samargandi OA, Tang DT. Variations in the anatomical structures of the Guyon canal. Plastic Surgery. 2017;**25**:84-92. DOI: 10.1177/2292550317694851

[51] Bozkurt MC, Tağil SM, Ozçakar L, Ersoy M, Tekdemir I. Anatomical variations as potential risk factors for ulnar tunnel syndrome: A cadaveric study. Clinical Anatomy. 2005;**18**:274-280. DOI: 10.1002/ca.20107

[52] Gross MS, Gelberman RH. The anatomy of the distal ulnar tunnel. Clinical Orthopaedics and Related Research. 1985;**196**:238-247

[53] Zeiss J, Guilliam-Haidet L. MR demonstration of anomalous muscles about the volar aspect of the wrist and forearm. Clinical Imaging. 1996;**20**:219-221. DOI: 10.1016/0899-7071(95)00013-5

Section 3

Mathematical Model of Contraction in Vascular Smooth Muscle

Chapter 6

Molecular Mechanisms and Targets of Cyclic Guanosine Monophosphate (cGMP) in Vascular Smooth Muscles

Aleš Fajmut

Abstract

Molecular mechanisms and targets of cyclic guanosine monophosphate (cGMP) accounting for vascular smooth muscles (VSM) contractility are reviewed. Mathematical models of five published mechanisms are presented, and four novel mechanisms are proposed. cGMP, which is primarily produced by the nitric oxide (NO) dependent soluble guanylate cyclase (sGC), activates cGMP-dependent protein kinase (PKG). The NO/cGMP/PKG signaling pathway targets are the mechanisms that regulate cytosolic calcium ($[Ca^{2+}]_i$) signaling and those implicated in the Ca^{2+}-desensitization of the contractile apparatus. In addition to previous mathematical models of cGMP-mediated molecular mechanisms targeting $[Ca^{2+}]_i$ regulation, such as large-conductance Ca^{2+}-activated K^+ channels (BKCa), Ca^{2+}-dependent Cl^- channels (ClCa), Na^+/Ca^{2+} exchanger (NCX), $Na^+/K^+/Cl^-$ cotransport (NKCC), and Na^+/K^+-ATPase (NKA), other four novel mechanisms are proposed here based on the existing but perhaps overlooked experimental results. These are the effects of cGMP on the sarco−/endo- plasmic reticulum Ca^{2+}-ATPase (SERCA), the plasma membrane Ca^{2+}-ATPase (PMCA), the inositol 1,4,5-trisphosphate (IP_3) receptor channels type 1 (IP_3R1), and on the myosin light chain phosphatase (MLCP), which is implicated in the Ca^{2+}-desensitization. Different modeling approaches are presented and discussed, and novel model descriptions are proposed.

Keywords: vascular smooth muscle, contraction, relaxation, nitric oxide, cyclic guanosine monophosphate, protein kinase G, Ca^{2+} signaling, desensitization, mathematical model, ionic fluxes

1. Introduction

Cyclic guanosine 3′,5′-monophosphate (cGMP) is an intracellular second-messenger that mediates a broad spectrum of physiologic processes in multiple cell types within the cardiovascular, gastrointestinal, urinary, reproductive, nervous, endocrine, and immune systems. In particular, cGMP signaling plays a vital role in the endothelium, vascular smooth muscle cells (VSMC), and cardiac myocytes. cGMP was first synthesized in 1960, and soon after, its endogenous production was detected in rats. In the late 70s, two separate experiments confirmed that the gas nitric oxide (NO) stimulated cGMP production by activating soluble guanylate

IntechOpen

cyclase (sGC). In 1980, it was reported that a diffusible substance causing vasodilatation is released from the endothelium. The so-called endothelium-derived relaxing factor (EDRF) was identified seven years later as NO. See [1] for review.

The molecular mechanisms of cardiovascular NO signaling are not entirely understood. Still, it is currently accepted that many effects are mediated, at least in part, via cGMP-dependent pathways. Within the cardiovascular system, these signaling pathways play a vital role in vasodilatation as well as in proliferation, migration, differentiation, and inflammation of VSMC and endothelial cells (ECs), in the modulation of myocyte contractility as well as of cardiac remodeling and thrombosis [2–4]. Impaired functioning at any signaling step from the synthesis through the effector activation and the degradation process of either NO or cGMP accounts for numerous cardiovascular diseases, such as hypertension, atherosclerosis, cardiac hypertrophy, and heart failure [3, 4]. Hence, these signaling pathways represent the potential targets for pharmacological treatment.

2. Nitric oxide (NO) and cyclic guanosine monophosphate (cGMP) production and degradation

Various stimuli can trigger relaxation responses of VSMC via the production and signaling of NO in the vascular endothelium. These are endogenous neurotransmitters (e.g., substance P and acetylcholine), humoral substances (e.g., bradykinin), and mechanical stimuli (e.g., the increase in hemodynamic shear stress or intraluminal pressure). They all trigger a complex cascade of biochemical reactions, accounting for either the mobilization, activation, or increase in the catalytic activity of NOS to produce NO or for upregulation of its gene expression. In the cardiovascular system, most of NO is produced in the endothelium by the endothelial NOS (eNOS). eNOS is also detected in cardiac myocytes, platelets, certain neurons, and kidney tubular epithelial cells. The other two isoforms are neuronal- and inducible- NOS (nNOS and iNOS, respectively). The former is mainly located in the nervous system, and the latter, which is induced by cytokines, is predominantly found in the immune system. They all catalyze the oxidation of the amino acid L-arginine into L-citrulline, where the by-product is NO [5].

Sensing of shear stress is still under intensive research since it is mediated by rapid and almost simultaneous activation of various membrane molecules and microdomains, including ion channels, tyrosine kinase receptors, G-protein-coupled receptors, caveolae, adhesion proteins, cytoskeleton, glycocalyx, primary cilia, and filaments [6]. Though the underlying biochemical signaling processes are not entirely understood, three main mechanisms of mechanotransduction were proposed. The first one involves the mechanisms which account for the entry of Ca^{2+} across the EC plasma membrane either via capacitive Ca^{2+} entry (CCE) [7] or via activation of mechanosensing ion channels (MSICs) [8]. Both processes lead to further increases in $[Ca^{2+}]_i$, its consequent interaction with calmodulin (CaM), and finally to NOS activation. The other two mechanisms cross-correlate many signaling pathways mediated by G protein-coupled receptors (GPCR) and integrins involving protein kinases A, B, C, and G (PKA, Akt, PKC, and PKG, respectively), as well as phosphatidylinositide 3-kinase (PI3K). These signaling pathways regulate the activation of different nuclear factors affecting NOS gene expression [9], the recruitment of NOS from caveolae, the phosphorylation of NOS, and the cytosolic $[Ca^{2+}]_i$ concentration and signaling [6].

Downstream the NO production cGMP is produced either by the soluble or the membrane-bound particulate guanylate cyclases, sGC and pGC, respectively, in response to either elevated NO or brain and atrial natriuretic peptides (BNP and

ANP, respectively). Natriuretic peptides (NPs) activate pGC, while NO diffuses into the cytosol, binds to, and activates sGC. cGMP exerts its action predominantly through binding and activating its target, cGMP-dependent protein kinase (PKG) [3]. There are two other types of cGMP-target effector molecules. The first type is phosphodiesterases (PDEs), which also degrade other cyclic nucleotides. The second type is nonselective cation channels, which are present in the visual and olfactory systems. PDEs degrade cGMP and, hence shape its spatiotemporal levels. CGMP also cross-regulates cyclic adenosine monophosphate levels (cAMP) since other PDEs (e.g. PDE2) that degrade both cAMP and cGMP are stimulated by cGMP [10]. In addition to PDE5, which selectively degrades cGMP, several other PDE isoforms can hydrolyze both, cGMP and cAMP. These are PDE1, PDE2, and PDE3. The strategy of inhibiting PDEs to enhance cGMP and related signaling has already been successfully used with the PDE5 inhibitors, especially sildenafil, to treat erectile dysfunction, pulmonary hypertension, and chronic heart failure [10]. Other cGMP-elevating drugs, such as nitrovasodilators that donate NO, and various NP analogs, have also been successfully used in humans to treat cardiovascular diseases. NO-generating drugs such as glyceryl trinitrate or sodium nitroprusside have been used to treat angina pectoris in humans for more than 100 years [11].

3. Calcium-contraction coupling in vascular smooth muscle cells (VSMC)

The contractile state of VSMCs is regulated dynamically by hormones and neurotransmitters via the increase of the cytosolic calcium concentration ($[Ca^{2+}]_i$). Ca^{2+} is mostly released from its intracellular store sarcoplasmic reticulum (SR) via IP_3 sensitive or ryanodine receptor channels (IP_3R and RyR, respectively). In part, Ca^{2+} entry to the cytosol could be ascribed to the fluxes across the plasma membrane via the Ca^{2+}- selective voltage-dependent channels. The rise in $[Ca^{2+}]_i$ initiates binding of Ca^{2+} to CaM and the consequent interactions of myosin light chain kinase (MLCK) with Ca^{2+}/CaM complexes. Active MLCK is the one that is bound with the Ca_4CaM complex. Active MLCK phosphorylates the regulatory myosin light chain (MLC), enabling the attachment of myosin heads to the actin filaments and cross-bridge cycling [12]. The smooth muscle cell's contractile state is determined by the extent of MLC phosphorylation regulated by the balance of MLCK and MLC phosphatase (MLCP) activities. The latter dephosphorylates MLC. High vascular tone is maintained as long as the phosphorylation rate is higher than that of dephosphorylation. Relaxation occurs when $[Ca^{2+}]_i$ decreases, which results in the dissociation of Ca^{2+} from CaM and inactivation of MLCK. In that case, the activity of MLCP predominates the activity of MLCK, and the active actin-myosin cross-bridge cycling is not established. However, a passive latch state is possible [13]. The level of smooth muscle contractility can also be modulated at constant $[Ca^{2+}]_i$. The protein kinase C (PKC) and Rho kinase (ROCK) pathways play an essential role in regulating MLCP activity. They may cause diminished activity of MLCP and result in increased levels of phosphorylated MLC and a higher tension at a given $[Ca^{2+}]_i$. This increased contractility is called Ca^{2+} sensitization [12]. In reality, the process is much more complex since it is composed of many cross-interacting pathways with different feedbacks, nonlinear behavior of the interactions, dynamical changes of many variables – especially $[Ca^{2+}]_i$ [14]. In this complex system of interactions $[Ca^{2+}]_i$ signaling still represents a bottleneck according to its bow-tie structure of encoding and decoding [15]. cGMP/PKG signaling occurs on both – the encoding and the decoding sides and represents a predominant mechanism in regulating vasoactivity, particularly vasorelaxation. More than ten substrates being phosphorylated *in vivo* by PKG were identified, and many of them take part in the $[Ca^{2+}]_i$ encoding and decoding processes [3, 16].

3.1 cGMP-dependent protein kinase (PKG)

The enzyme PKG belongs to the family of serine/threonine (Ser/Thr) kinases. In mammals, PKG-I and PKG-II are encoded by different genes, prkg1 and prkg2, respectively. PKG-I exists in two isoforms PKG-Iα and PKG-Iβ. PKG-I is present at high concentrations in all smooth muscles, including the uterus, vessels, intestine, and trachea. PKG-II is expressed in several brain nuclei, intestinal mucosa, kidney, adrenal cortex, chondrocytes, and lung. Only PKG-Iα and PKG-Iβ are expressed in the vascular system. See [3] for a review. All types of PKG are homodimers. Each monomer contains a regulatory and a catalytic domain. Each of the PKG regulatory domains binds two cGMP molecules allosterically with high cooperativity. The affinities of the two binding sites on each of the subunits of PKG-Iα differ by approximately tenfold. Binding sites occupied by cGMP induce significant conformation changes in the molecular structure. By that, autoinhibition of the catalytic center is released, and the basal activity is increased. Hence, the phosphorylation of serine/ threonine residues of the target proteins as well as of the autophosphorylation site is possible. All four binding sites have to be occupied with cGMP for the fully active holoenzyme PKG [17]. PKG-I mediates both receptor-triggered and depolarization-induced vasorelaxation by several mechanisms. Many of them are not entirely understood, and some of them are still unknown. In general, PKG-mediated relaxation is induced either by attenuation of $[Ca^{2+}]_i$ and/or desensitization of the contractile apparatus. The first effect is achieved by negatively affecting the "$[Ca^{2+}]_i$-on" mechanisms and by positively affecting the "$[Ca^{2+}]_i$-off" mechanisms. On the $[Ca^{2+}]_i$-decoding part, PKG's effect is concentrated mainly on the activation of the MLCP, which desensitizes the contractile apparatus to $[Ca^{2+}]_i$ [18].

3.2 "$[Ca^{2+}]_i$-on" mechanisms as targets of cGMP/PKG signaling

One of the primary targets of cGMP/PKG signaling to elevate $[Ca^{2+}]_i$ is the IP_3 receptor channel type 1 (IP_3R1) and its correlated cGMP-associated kinase substrate protein (IRAG). If IRAG is colocalized with IP_3R1 and PKG-Iβ in the presence of cGMP, it inhibits Ca^{2+} release through IP_3R1 via its phosphorylation [19]. It was shown that PKG-Iβ exclusively phosphorylated only the type 1 but not the type 2 and 3 IP_3R *in vivo* and that both, PKG and PKA, phosphorylated IP_3R1 *in vitro* in gastric smooth muscles, which resulted in diminished IP_3 and Ca^{2+}-induced Ca^{2+} release (ICICR) from the SR [20]. *In vivo* experiments on mice with mutated IRAG, which did not interact with IP_3R, showed that PKG/IRAG/ IP_3R interactions indeed decrease the receptor-triggered $[Ca^{2+}]_i$ and hence contraction [21]. In the *in vitro* experiments, it was also shown that PKG-Iβ phosphorylated IRAG but not IP_3R [22]. The same was confirmed with COS-7 transfected cells where the phosphorylation of IRAG resulted in the reduced Ca^{2+} release during concurrent activation of PKA and PKG. The effect was observed for all three IP_3R sub-types [23]. It is supposed that IRAG signaling does not modulate basal tone but might be important for blood pressure regulation under pathophysiological conditions [24].

PKG-Iα may also attenuate receptor-activated contraction via inhibition of IP_3 production mediated by GPCR signaling [25] and interfering with phospholipase C-β (PLC-β) [26]. It has been shown that the isoform PKG-Iα binds, phosphorylates, and activates the regulator of G protein signaling 2 (RGS2), which terminates the signal transduction of the contractile agonists mediated by the Gq-coupled receptors and terminates thereby the activity of PLC [25]. It was also proposed that PKG-Iα/RGS2 pathway might inhibit hormone receptor-triggered Ca^{2+} release and vasoconstriction *in vivo* [27]. It has also been shown that PKG can directly phosphorylate PLC-β *in vitro* in cultured COS-7 cells and *in vivo* in aortic VSMC, which blocked the activation of the enzymes correlated with the G-protein subunits and attenuated agonist-induced IP_3 production and Ca^{2+} release [26].

There is also evidence that PKG may cause vasodilatation by suppressing the Ca^{2+} influx across the plasma membrane through the voltage-operated Ca^{2+} channels (VOCC). cGMP/PKG has the opposite effect as cAMP/PKA on this type of channel. The former inhibited and the latter enhanced L-type Ca^{2+} channel (LTCC) activity in rabbit portal vein myocytes [28]. On the other hand, in rat cerebral arterial VSMC, which express T-type Ca^{2+} channels (TTCC) PKA [29] and PKG [30] both had a suppressing effect on their conductance. In both cases, a rightward shift of the voltage-response curve was observed. A similar effect was observed for the nonselective transient receptor potential cationic 1/3 channels (TRPC1/3) [31]. On the other hand, the experiments on the macroscopic and single-channel Ca^{2+} currents from guinea-pig basilar artery showed that the addition of 10 μM cGMP did not affect single-channel properties, such as conductance, voltage dependence, the number of open states, and different time constants, but significantly reduced the channel availability [32].

3.3 "$[Ca^{2+}]_i$-off" mechanisms as targets of cGMP/PKG signaling

cGMP/PKG is supposed to enhance the activities of all three major Ca^{2+}-removal systems in VSMCs. The Primary $[Ca^{2+}]_i$-off mechanism is refilling the Ca^{2+} stores via sarco−/endo- plasmic reticulum Ca^{2+}-ATPase (SERCA). The increase in SERCA activity in response to cGMP was first identified in isolated SR vesicles from cardiac and smooth muscles [33]. Later it was demonstrated that NO-induced relaxation of cultured VSMC from the aorta was associated with increased PKG-dependent phospholamban (PLB) phosphorylation [34]. Using a solid-state nuclear magnetic resonance (NMR) spectroscopy, it was found that PLB binds to SERCA allosterically [35]. Moreover, the phosphorylation at Ser16 of PLB, which gradually lowers PLB interaction with SERCA, was found to increase SERCA activity [35]. In gastric SMC, cGMP-mediated Ca^{2+} uptake via SERCA was observed *in vitro* in a concentration-dependent manner [36].

Experiments on cultured aortic VSMC provided evidence that cGMP also accelerates $[Ca^{2+}]_i$ extrusion by stimulating the Na^+/ Ca^{2+} exchangers (NCX) at different Na^+ concentrations [37]. cGMP increased both forward and reversed Na^+/ Ca^{2+} exchange modes by approximately 50% after adding 500 μM of membrane-permeable cGMP analog. The $[Ca^{2+}]_i$ pumping activity gradually increased with cGMP concentration. Phosphorylation by PKG was proposed as the underlying mechanism for this effect [37].

Another cGMP/PKG-mediated $[Ca^{2+}]_i$-off mechanism is the plasma membrane Ca^{2+} ATP-ase (PMCA). The evidence was first obtained with experiments on isolated proteins [38] and experiments performed on cultured VSMC [39]. All results suggested that the phosphorylation of the PMCA by PKG was responsible for stimulating the Ca^{2+}-pumping activity, which was 2.4-fold higher after adding 500 μM of membrane-permeable cGMP analog. The leftward shift in the pumping activity vs. $[Ca^{2+}]_i$ dependence was also observed [39]. Experiment on isolated and purified PMCA from porcine aorta [40] confirmed the previous two results at much smaller cGMP concentrations.

3.4 cGMP/PKG-dependent mechanisms that indirectly affect $[Ca^{2+}]_i$

The mechanisms by which cGMP/PKG signaling interferes with $[Ca^{2+}]_i$ are primarily linked with cell-membrane depolarization/hyperpolarization. Although depolarization-induced contraction remains mostly unresolved, these mechanisms are intensively studied [41]. One of the established targets of cGMP/PKG signaling is the large-conductance Ca^{2+}-activated K^+ channel (BKCa). The modulation of BKCa by different protein kinases in different smooth muscle tissues as well as the sites and mechanisms of their action remain unresolved [42]. The activation BKCa presumably hyperpolarises the cell membrane, thereby influences the gating of voltage-operated Ca^{2+} channels and lowers $[Ca^{2+}]_i$. PKG-I is known to activate BKCa either directly by phosphorylation [43] or indirectly via protein phosphatase regulation [44]. Activation of BKCa in the presence of NO/cGMP in isolated rat afferent arterioles attenuated extracellular Ca^{2+} influx upon KCl stimulation [45]. The role and importance of BKCa in vasorelaxation were highlighted with the experiments performed on BKCa-deficient mice. Their deletion led to a relatively mild increase in blood pressure. However, it increased vascular tone in small arteries due to a complete lack of spontaneous K^+ efflux and, therefore, depolarised state of the membrane, and reduced suppression of Ca^{2+} transients in response to cGMP [46].

Another mechanism by which cGMP/PKG signaling may affect $[Ca^{2+}]_i$ influx is via Ca^{2+}-activated Cl^- channels [47]. These type of channels was observed in VSMC of mesenteric resistance arteries. Since their activation required phosphorylation, was sensitive to PKG inhibitors, and was evoked by adding PKG, it is believed that the effect of cGMP on the Cl^- current is mediated through PKG [47]. The physiological role of Ca^{2+}-activated Cl^- channels is ambiguous since their excessive activation would promote an inward Cl^- current leading to cell depolarization, activation of VOCC, increase in $[Ca^{2+}]_i$, and, hence, vasoconstriction.

Information on the effect of cGMP/PKG on Na^+/K^+ ATPase (NKA) [48] and cotransport of $Na^+/K^+/Cl^-$ (NKCC) [49] in terms of VSMC physiology is very limited and vague. However, these mechanisms have been implicated in the mathematical models [50, 51]. It was reported that cGMP might increase the activity of NKCC in vascular SMC of rat thoracic aorta by up to 3.5-fold [49]. In the canine pulmonary artery SMC, nitroprusside/cGMP-mediated relaxation was accompanied by increase NKA activity [48].

3.5 cGMP/PKG signaling targeting the Ca^{2+}-desensitization mechanisms

PKG may also cause vasodilatation by desensitizing the contractile apparatus in response to elevated $[Ca^{2+}]_i$, resulting from either MLCP activation or MLCK deactivation. Both effects lead to MLC dephosphorylation, myosin cross-bridge detachment, and relaxation even at high $[Ca^{2+}]_i$. The enzyme MLCP plays a major role in the Ca^{2+}-desensitization since it is not directly Ca^{2+}-dependent, and it embodies various possibilities for regulating its activity [52]. These different options arise from its complex structure and widespread distribution in different tissues. MLCP holoenzyme is composed of three subunits – catalytic (PP1c), regulatory (MYPT1), and a small subunit (M20/M21). It is a Ser/Thr phosphatase that belongs to the protein phosphatase type 1 (PP1) family. Active PP1c is required for its catalytic activity, while MYPT1 targets the enzyme to its substrates and also autoregulates the catalytic activity of PP1c. This autoregulation emerges because MYPT1 contains different, for its structure and activity important, phosphorylation sites. In human sequence, these phosphorylation sites are Thr696 and Thr853, which are phosphorylated by ROCK [53] and other agonist-induced kinases. There are also Ser695 and Ser852 phosphorylation sites on MYPT1, which are phosphorylated by PKA and

PKG [54]. The residues Ser695 and Thr696 as well as Ser852 and Thr853, are close within the MYPT1 sequence, and thus phosphorylation of one site prevents the phosphorylation of the neighboring site. It was proposed and also demonstrated that PKA or PKG-dependent phosphorylation of Ser695 and Ser852 prevents the phosphorylation of Thr696 and Thr853 and vice versa [12, 54, 55].

The current hypothesis is that the phosphorylation of Thr696 and Thr853 induces such structural changes in MYPT1 that these phosphorylated sites interact with the MLCP catalytic subunit PP1c [56] and is supported by the fact that MYPT1 is quite flexible at this part of the structure. Moreover, the sequences around Thr696 or Thr853 are similar to that of Ser19, where MLC is phosphorylated [57]. It is hypothesized that P-Thr696 and P-Thr853 may represent either substrate analogs to P-Ser19 of MLC or a potent autoinhibitory site docking to the PP1c catalytic subunit of MLCP [56]. In all these scenarios, the MLCP-dependent rate of MLC dephosphorylation is decreased. On the other hand, if MYPT is phosphorylated at Ser695 and Ser852 beforehand, Thr696 and Thr853 phosphorylation is blocked [54, 56].

Phosphorylation of Thr853 is a less potent inhibitor of MLCP than Thr696 [56]. It was also reported that PKA could phosphorylate all four sites, Ser695, Thr696, Ser852, Thr853, simultaneously. However, such a form of MYPT1 did not inhibit PP1c [58]. Another possibility of MLCP activity inhibition is binding the phosphorylated form of PKC-potentiated phosphatase inhibitor protein of 17 kDa (CPI-17) to the catalytic subunit PP1c. The phosphorylation increases the affinity of CPI-17 for PP1c by approximately 1000-fold, resulting in suppressed MLCP activity [59]. CPI-17 is expressed predominantly in tonic smooth muscles with slow and sustained contraction, especially in VSMC from the aorta and femoral arteries. The enzymes linked with the phosphorylation of CPI-17 are PKC, ROCK, zipper-interacting protein kinase (ZIPK), integrin-linked kinase (ILK). However, PKC and ROCK are most commonly mentioned [60]. ROCK signaling interferes with PKG and PKA signaling since PKA and PKG phosphorylate RhoA, the ROCK activator. Increased level of RhoA phosphorylation attenuates ROCK activity. In this way, PKG mediates vasorelaxation via reduced activity of ROCK and the correlated reduced inhibition of MLCP. That leads to faster MLC dephosphorylation and relaxation [61].

The role of PKC and ROCK in the stimulation-contraction coupling is still not well understood [62]. It is also possible that their role and importance in different smooth muscles is different. However, it is believed that CPI-17 phosphorylation and the corresponding inhibition of MLCP is the predominant process of the early phase of contraction. It was reported that PKC is believed to be primarily responsible for fast CPI-17 phosphorylation during the early phase of vasoconstriction, and ROCK was found responsible for slow, sustained CPI-17 phosphorylation during the sustained phase of contraction [63]. On the other hand, in the rat airways, ROCK activation and the consequent MLCP inhibition contributed to the early phase of the smooth muscles' contractile response. Whatever the agonist in that system was, the ROCK inhibitor Y27632 did not modify the basal tension. Still, it decreased the amplitude of the short duration response without altering the superimposed delayed contraction [64]. That indicates that in rat airway SMC, ROCK plays a major role in CPI-17 phosphorylation and that other kinases are responsible for Thr696 and Thr853 phosphorylation [62].

Moreover, PKG may affect MLCP activity also by the phosphorylation of telokin, which is a smooth muscle-specific protein whose sequence is identical to that of the noncatalytic terminus of MLCK. Telokin does not increase MLCP activity *per se* but acts synergistically with PKA and PKG [65]. By binding to either phosphorylated MYPT1 and/or phosphorylated MLC, telokin is supposed to facilitate the interaction between the enzyme and its substrate and de-inhibits the auto-

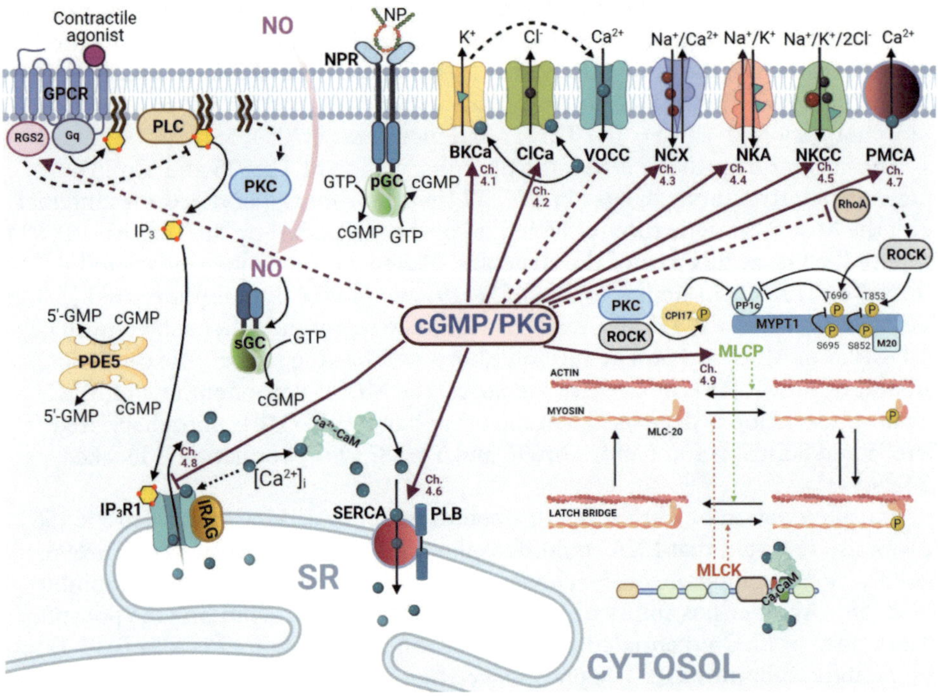

Figure 1.
Molecular mechanisms and targets of cyclic guanosine monophosphate (cGMP)/protein kinase G (PKG) signaling in vascular smooth muscle cells (VSMCs) described in chapters 4.1 to 4.9 (full dark red lines denoted with Ch. 4.1 to 4.9) and others described in the text (dashed dark red lines). For explanation see text. Abbreviations used: GPCR (G protein-coupled receptor), RGS2 (regulator of G_q protein signaling 2), G_q (G protein), PLC (phospholipase C), PKC (protein kinase C), IP_3 (inositol 1,4,5-trisphosphate), NO (nitric oxide), NPR (natriuretic peptide receptor), NP (natriuretic peptide), GTP (guanosine 5′-triphosphate), cGMP (cyclic guanosine monophosphate), 5′-GMP (guanosine 5′-monophosphate), pGC (particulate guanylate cyclase), sGC (soluble guanylate cyclase), PDE5 (phosphodiesterase 5), BKCa (large-conductance Ca^{2+}-activated K^+ channels), ClCa (Ca^{2+}-dependent Cl^- channels), VOCC (voltage-operated Ca^{2+} channel), NCX (Na^+/Ca^{2+} exchanger), NKA (Na^+/K^+-ATPase), NKCC ($Na^+/K^+/Cl^-$ cotransport), PMCA (plasma membrane Ca^{2+}-ATPase), RhoA (ROCK activator), ROCK (rho kinase), CPI-17 (PKC-potentiated phosphatase inhibitor protein of 17 kDa), PP1c (MLCP catalytic subunit), MYPT1 (MLCP regulatory subunit), M20 (MLCP small subunit), MLCP (myosin light chain phosphatase), T696/ T853 (threonine 696/ 853 of the MYPT1), S695/S852 (serine 696/853 of the MYPT1), MLCK (myosin light chain kinase), MLC-20 (20 kDa myosin light chain), Ca_4CaM (calmodulin with bound 4 Ca^{2+}), Ca^{2+}-CaM (Ca^{2+}-calmodulin complexes), SR (sarcoplasmic reticulum), SERCA (sarco–/endo- plasmic reticulum Ca^{2+}-ATPase), PLB (phospholamban), $[Ca^{2+}]_i$ (cytosolic Ca^{2+} concentration), IRAG (IP_3R1-correlated cGMP-associated kinase substrate protein), IP_3R1 (1,4,5-trisphosphate receptor channel type 1), P (phosphorylated form of protein).

suppressed MLCP activity emerging from Thr696 and Thr853 phosphorylation. This mechanism results in an increased rate of MLC dephosphorylation [66]. The majority of the described mechanisms and targets of cGMP/PKG signaling are summarized in **Figure 1**. 11 targets are depicted, from which 9 of them are described by mathematical models presented in the following chapter.

4. Mathematical modeling of cGMP/PKG-mediated ionic fluxes and Ca^{2+}-desensitization of the contractile apparatus

The first attempt to build a whole-cell-like model of VSMC, also considering the NO/sGC/cGMP signaling cascade, was performed by Yang et al. [67]. They upgraded their existing models for rat cerebrovascular arteries [68]. The model [67] predicted the NO-induced cGMP production and the corresponding attenuation of

$[Ca^{2+}]_i$, Ca^{2+}-desensitization of the contractile apparatus, and the reduction in force. In terms of cGMP-mediated target-regulation, they considered the effects on the BKCa and the contractile mechanism. Model simulations reproduced major NO/cGMP-induced VSMC relaxation effects. Additionally, cGMP was also considered in sGC desensitization, limiting cGMP production well below maximum [67]. The activating effect of NO/cGMP on BKCa was assumed as cGMP-dependent and partially NO-dependent.

In 2007 and 2008, another two whole-cell-like models for VSMC were presented [50, 51]. However, both focused only on $[Ca^{2+}]_i$ signaling and did not consider the processes of the contractile apparatus. Jacobsen et al. [50] focused primarily on the role of Ca^{2+}-dependent Cl^- channels that may cause the transitions between different types of $[Ca^{2+}]_i$ signals in rat mesenteric small arteries upon α-adrenoreceptor stimulation. Instead of cGMP's influence on BKCa, they considered the cGMP-dependent mechanisms of Ca^{2+}-dependent Cl^- channels and NKA. Kapela et al. [51] focused primarily on the plasma membrane electrophysiological properties and considered eleven ionic currents across the plasma membrane. Four of them considered cGMP-dependent mechanisms, i.e. Ca^{2+}-dependent Cl^- channels, BKCa, NCX, and NKCC. The model's purpose was to provide a working database of the rat mesenteric SMC physiological data. It was considered as the building block of the future multi-cellular models of the vascular wall [51].

4.1 The model of cGMP-mediated current through the large-conductance Ca^{2+}-activated K^+ channels (BKCa)

BKCa is the most frequently modeled cGMP-dependent mechanism accounting for $[Ca^{2+}]_i$ signaling. The first model of Yang et al. [67] was based on the experimental studies of Zhou et al. [69], who suggested that PKG stimulates the activity of two isoforms of BKCa either by phosphorylation of the channel or its regulatory proteins. The resultant effect on the potassium electric current (I_K) was modeled with a left-shift of the voltage dependency of equilibrium open probability ($\overline{P}_{K,o}$) towards more negative potentials [67]. The complete mathematical description of [68] follows the Hodgkin-Huxley formalism. The general expression for I_K is:

$$I_K = g_K P_{K,o}(V_m - V_K), \tag{1}$$

where g_K is a channel conductance, $P_{K,o}$ is open probability or gating of the channel and $(V_m - V_K)$ is the driving force of the current, where V_m is a membrane potential, and the V_K is the Nernst equilibrium potential. Eq. (1) is analogous to Ohm's law. The overall gating factor $P_{K,o}$ consists of two parts – a fast gating term ($P_{K,f}$) and a slow gating term ($P_{K,s}$):

$$P_{K,o} = f_K P_{K,f} + s_K P_{K,s}, \tag{2}$$

where f_K is a fraction of fast channels, and s_K is a fraction of slow channels. Fast and slow gating terms are described with a first-order ordinary differential equation for a biphasic (open-close) transition:

$$\frac{dP_{K,f}}{dt} = \frac{\overline{P}_{K,o} - P_{K,f}}{\tau_{K,f}}, \tag{3}$$

$$\frac{dP_{K,s}}{dt} = \frac{\overline{P}_{K,o} - P_{K,s}}{\tau_{K,s}}, \tag{4}$$

where $\tau_{K,f}$ and $\tau_{K,s}$ are the characteristic opening times and $\overline{P}_{K,o}$is an equilibrium open probability, which is a sigmoidal function of the membrane potential (V_m):

$$\overline{P}_{K,o} = \frac{1}{1 + e^{-\left(V_m - V_{K,1/2}\right)/S_{K,0}}}. \tag{5}$$

The parameter's value $S_{0,K}$ represents the slope of the sigmoidal function, and its sign defines the orientation (declining/ increasing). Typically, $V_{1/2,K}$ is a parameter and represents the membrane potential, at which half-maximal value $\overline{P}_{K,o}$ is achieved; however, here, it is a function of $[Ca^{2+}]_i$, $[cGMP]$, and $[NO]$:

$$V_{K,1/2} = -V_{K,Ca} \log\left(\left[Ca^{2+}\right]_i\right) - V_{K,0} - V_{K,cGMP}R_{K,cGMP} - V_{K,NO}R_{K,NO}, \tag{6}$$

where $V_{K,0}$ is a basal value of $V_{1/2}$, and $V_{K,NO}$, $V_{K,cGMP}$ and $V_{K,Ca}$ are maximal induced shifts of $V_{1/2}$ towards lower values, and, $R_{K,cGMP}$ and $R_{K,NO}$ are the regulatory Hill functions:

$$R_{K,cGMP} = \frac{[cGMP]^{n_{cGMP,K}}}{[cGMP]^{n_{cGMP,K}} + K_{cGMP,K}^{n_{cGMP,K}}}, \tag{7}$$

$$R_{K,NO} = \frac{[NO]^{n_{NO,K}}}{[NO]^{n_{NO,K}} + K_{NO,K}^{n_{NO,K}}}, \tag{8}$$

where $n_{cGMP,K}$ and $n_{NO,K}$ are the Hill coefficients and, $K_{cGMP,K}$ and $K_{NO,K}$ are the half-saturation constants. The same notation for Hill function parameters is used elsewhere in the text. The descriptions of all parameters are given in tables.

Authors Kapela et al. [51] used almost the same approach as Yang et al. [67]. In the former cas the authors used the Goldman-Hodgkin-Katz model to describe the potassium flux I_K:

$$I_K = A_m N_{BKCa} P_{K,o} P_{BKCa} V_m \frac{F^2}{RT} \frac{[K]_o - [K]_i e^{\frac{F}{RT}Vm}}{1 - e^{\frac{F}{RT}Vm}}, \tag{9}$$

where A_m is a cell-membrane surface area, N_{BKCa} is a channel density, P_{BKCa} is a single channel permeability, V_m is a membrane potential, $[K]_o$ and $[K]_i$ are the external and internal potassium concentrations, respectively, F is a Faraday constant, R is the universal gas constant, and T is the absolute temperature. These cell-specific and general parameter values could be found in [51]. cGMP-dependent gating $P_{K,o}$ is defined the same as above in Eqs. (2)–(8).

The comparison of parameter values presented in **Table 1** reveals similarities but also differences. The model of Kapela et al. [51] was written more specifically for the rat mesenteric arteriole, whereby the parameters for the BKCa were such that they fitted experimental data of [70]. In contrast, the model of Yang et al. [67] was compared with the experimental data for rabbit femoral arteries [71], and the parameters for BKca accounted for [72].

4.2 The model of cGMP-mediated current through the Ca^{2+} activated Cl^- channels (ClCa)

The model was first proposed by Jacobsen et al. [50] and was based on the measurements performed on the rat mesenteric resistance arteries [47].

Parameter	Description	Value [67]	Value [51]
g_K	Overall maximal conductance of the BKCa	0.5 nS	/
f_K	Fraction of fast channels	0.65	0.17
s_K	Fraction of slow channels	0.35	0.83
$\tau_{K,f}$	The characteristic time of fast-channel activation	0.5 ms	0.84 ms
$\tau_{K,s}$	The characteristic time of slow-channel activation	11.5 ms	35.9 ms
$S_{K,0}$	The slope of the $\overline{P}_o$ vs. V_m function	30.8 mV	18.25 mV
$V_{K,Ca}$	Maximal Ca^{2+}-induced $V_{1/2}$ shift	53.7 mV	41.7 mV
$V_{K,0}$	Basal $V_{1/2}$value	283.7 mV	128.2 mV
$V_{K,cGMP}$	Maximal cGMP-induced $V_{1/2}$ shift	66.9 mV	76 mV
$V_{K,NO}$	Maximal NO-induced $V_{1/2}$ shift	100 mV	46.3 mV
$n_{cGMP,K}$	Hill coefficient	2	2
$n_{NO,K}$	Hill coefficient	1	1
$K_{cGMP,K}$	Half saturation constant in the regulatory cGMP-dependent Hill function	0.55 μM	1.5 μM
$K_{NO,K}$	Half saturation constant in the regulatory NO-dependent Hill function	0.2 μM	0.2 μM

Table 1.
Parameter values for the cGMP-dependent Ca^{2+}-activated K^+ (BKCa) current.

The Cl^- electric current (I_{Cl}) across the plasma membrane is defined as that for potassium in Eq. (1):

$$I_{Cl} = g_{Cl} P_{Cl,o}(V_m - V_{Cl}), \tag{10}$$

The expression for $P_{Cl,o}$ is analogous to Eq. (3):

$$\frac{dP_{Cl,o}}{dt} = \frac{\overline{P}_{Cl,o} - P_{Cl,o}}{\tau_{Cl}}, \tag{11}$$

but the equilibrium open probability $\overline{P}_{Cl,o}$ is not defined according to the Hodgkin-Huxley formalism but rather with an adapted Hill-type function:

$$\overline{P}_{Cl,o} = R_{cGMP,Cl} \frac{\left[Ca^{2+}\right]_i^{n_{Ca,Cl}}}{\left[Ca^{2+}\right]_i^{n_{Ca,Cl}} + \left(K_{Ca,cGMP,Cl}(1 - \rho R_{cGMP,Cl})\right)^{n_{Ca,Cl}}}, \tag{12}$$

where $n_{Ca,Cl}$, $K_{Ca,cGMP,Cl}$ and ρ are parameters, and $R_{cGMP,Cl}$ is cGMP-dependent:

$$R_{cGMP,Cl} = \frac{[cGMP]^{n_{cGMP,Cl}}}{[cGMP]^{n_{cGMP,Cl}} + K_{cGMP,Cl}^{n_{cGMP,Cl}}}. \tag{13}$$

For I_{Cl}, Kapela et al. [51] used a similar approach as Jacobsen et al. [50]. However, the former authors applied slight modifications. The general description of the Cl^- current is the same as in [50] (Eq. (10)), only the expression $P_{Cl,o}$ is different. Kapela et al. [51] considered that a fraction of Ca^{2+}-dependent Cl^- channels is cGMP dependent, and a fraction is cGMP-independent. That is evident from the two terms within the expression for $P_{Cl,o}$:

$$P_{Cl,o} = R_{I,Cl}\frac{\left[Ca^{2+}\right]_i^{n_{Ca,Cl}}}{\left[Ca^{2+}\right]_i^{n_{Ca,Cl}} + K_{Ca,Cl}^{n_{Ca,Cl}}} + R_{cGMP,Cl}\frac{\left[Ca^{2+}\right]_i^{n_{Ca,Cl}}}{\left[Ca^{2+}\right]_i^{n_{Ca,Cl}} + \left(K_{Ca,cGMP,Cl}\left(1-\rho R_{cGMP,Cl}\right)\right)^{n_{Ca,Cl}}}, \quad (14)$$

where $R_{I,Cl}$ is a cGMP-independent component and $R_{cGMP,Cl}$ is defined the same as in Eq. (13). Parameter values and their descriptions are presented in **Table 2**.

The comparison of parameter values presented in **Table 2** shows remarkable similarity. However, there are two significant differences in the modeling approach. Jacobsen et al. [50], who first proposed the cGMP-dependent model for I_{Cl}, defined $P_{Cl,o}$ as a time-dependent function, whereas Kapela et al. [51] proposed an equilibrium model and omitted differential Eq. (11). On the other hand, they added a cGMP-independent term (compare Eq. (12) and Eq. (14)). In both cases, the same reference with experimental data for rat mesenteric arteries was used to determine the parameter values [47], except for the half-saturation constant in the cGMP-independent term ($K_{Ca,Cl}$) of [51], which was determined for the rat portal vein SMC [73].

It is suggested that the effect of cGMP on Ca^{2+}-activated Cl^- current is not likely to be essential for the tonic receptor-activated contractile response but rather for the synchronization among VSMCs as between VSMCs and ECs [47, 50, 74].

4.3 The model of cGMP-mediated current through the Na^+/Ca^{2+} exchanger (NCX)

The framework for the mathematical description of the plasma membrane Na^+/Ca^{2+} exchange (NCX) (I_{NCX}) in Kapela et al. was taken from the model of Di Francesco and Noble 1985 [75], which was developed for the atrial myocytes. Kapela et al. [51] adjusted the maximal exchanger conductivity, which is much lower in SMC than in atrial myocytes, and added the effect of cGMP according to the measured results of [37]:

$$I_{NCX} = I_{NCX,s}R_{NCX,cGMP}\frac{\left[Na^+\right]_i^3\left[Ca^{2+}\right]_o e^{\frac{\gamma V_m F}{RT}} - \left[Na^+\right]_o^3\left[Ca^{2+}\right]_i e^{\frac{(\gamma-1)V_m F}{RT}}}{1 + d_{NCX}\left(\left[Na^+\right]_o^3\left[Ca^{2+}\right]_i - \left[Na^+\right]_i^3\left[Ca^{2+}\right]_o\right)}, \quad (15)$$

Parameter	Description	Values [50]	Values [51]
g_{Cl}	Overall maximal conductance	3.8 nS	5.75 nS
τ_{Cl}	The characteristic time constant of channel activation	50 ms	/
$n_{Ca,Cl}$	Hill coefficient	3	2
$K_{Ca,cGMP,Cl}$	Half saturation constant in cGMP-dependent factor	0.4 μM	0.4 μM
ρ	The determinant of cGMP influence on the half-saturation constant	0.9	0.9
$n_{cGMP,Cl}$	Hill coefficient	3.3	3.3
$K_{cGMP,Cl}$	Half saturation constant in the regulatory cGMP-dependent Hill function	6.4 μM	6.4 μM
$K_{Ca,Cl}$	Half saturation constant in cGMP-independent term	/	0.365 μM
$R_{I,Cl}$	Weight of the cGMP-independent term	/	0.0132

Table 2.
Parameter values for the cGMP-dependent Ca^{2+}-activated Cl^- current.

where $I_{NCX,s}$ is a scaling factor for I_{NCX}, d_{NCX} is the denominator constant, γ is voltage-dependence parameter, and $R_{NCX,cGMP}$ is a cGMP-dependent regulatory function:

$$R_{NCX,cGMP} = 1 + f_{NCX,cGMP} \frac{[cGMP]}{[cGMP] + K_{cGMP,NCX}}, \tag{16}$$

where $f_{NCX,cGMP}$ is an additional fold-increase in NCX current due to cGMP and $K_{cGMP,NCX}$ is a half-saturation constant. Parameter values are presented in **Table 3**.

In experiments [37], $[Ca^{2+}]_i$ pumping activity gradually increased with cGMP concentration. However, a 50% increase in Na^+/ Ca^{2+} exchange was observed after adding a large, probably unphysiological concentration (500 μM) of membrane-permeable cGMP analog. Hence, the effects of low cGMP concentrations on the overall $[Ca^{2+}]_i$ and contractile response are expected to be small. That is also evident from a large half-saturation constant ($K_{cGMP,NCX}$) in the cGMP-dependent function $R_{NCX,cGMP}$. However, the overall effect should be tested by integrating all mechanisms in a whole-cell-like VSMC model.

4.4 The model of cGMP-mediated current through the Na^+/K^+-ATPase (NKA)

The NKA pumps Na^+ out and K^+ in and has stoichiometry 3 Na^+:2 K^+. Jacobsen et al. [50] modeled the whole-cell electric current through NKA (I_{NaK}) as in [68]:

$$I_{NaK} = I_{NaK,max} \frac{[K^+]_i}{[K^+]_i + K_{K,NaK}} \frac{[Na^+]_i^{n_{NaK,Na}}}{[Na^+]_i^{n_{NaK,Na}} + K_{Na,NaK}^{n_{NaK,Na}}} \frac{V_m + \Delta V_1}{V_m + \Delta V_2}, \tag{17}$$

whereby the maximal current ($I_{NaK,max}$) is considered as linearly cGMP-dependent:

$$I_{NaK,max} = k_{1,NaK,cGMP}[cGMP] - k_{2,NaK,cGMP}. \tag{18}$$

All parameter descriptions and their values are presented in **Table 4**.

In terms of membrane potential, increased NKA activity hyperpolarizes the membrane and enhances the Ca^{2+} influx through VOCC, which is similar to the effect of cGMP on BKCa. The effect of cGMP/PKG on NKA has not been studied often. The mathematical model is built on a single measurement on purified pig renal NKA at one single concentration of cGMP, which in addition to PKG increased the activity 1.6-fold. cGMP alone did not change the activity, and PKG alone increased it 1.2-fold [76]. Due to the lack of credible measurements, the reliability of this model is limited.

Parameter	Description	Values [51]
$I_{NCX,s}$	Current scaling factor	0.0487–0.487 pA
d_{NCX}	Denominator constant	3×10^{-4}
γ	Voltage-dependence parameter	0.45
$f_{NCX,cGMP}$	Additional fold increase in electric current due to cGMP	0.55
$K_{cGMP,NCX}$	Half saturation constant in cGMP-dependent term	45 μM

Table 3.
Parameter values for the cGMP-dependent Na^+/ Ca^{2+} exchange (NCX) current.

Parameter	Description	Values [50]
$K_{K,NaK}$	Half-saturation constant	1 mM
$K_{Na,NaK}$	Half-saturation constant	11 mM
$n_{NaK,Na}$	Hill coefficient	1.5
ΔV_1	Electric potential shift	150 mV
ΔV_2	Electric potential shift	200 mV
$k_{1,NaK,cGMP}$	cGMP-concentration weighted electric current	30 pA/μM
$k_{2,NaK,cGMP}$	Electric current constant	30 pA

Table 4.
Parameter values for the cGMP-dependent current trough Na^+/K^+-ATPase (NKA).

4.5 The model of cGMP-mediated current through the $Na^+/K^+/Cl^-$ cotransporter (NKCC)

Instead of cGMP-dependent NKA, Kapela et al. [51] modeled the cGMP influence on the $Na^+/K^+/Cl^-$ cotransport (NKCC) having the 1:1:2 stoichiometry. The expression describing the electric current for a particular ion (I^i_{NaKCl}, where i is either Na, K, or Cl) was taken from [77] and upgraded with a cGMP dependency. According to [77], electric currents of individual ions are defined according to the valence (Z) and the stoichiometry:

$$I^{Na}_{NaKCl} = I^{K}_{NaKCl} = -\frac{1}{2} I^{Cl}_{NaKCl}. \tag{19}$$

Here only the electric current for Cl^- (I^{Cl}_{NaKCl}) is written:

$$I^{Cl}_{NaKCl} = -I_{NaKCl} Z_{Cl} R_{NaKCl,cGMP} \ln\left(\frac{[Na^+]_o}{[Na^+]_i}\frac{[K^+]_o}{[K^+]_i}\left(\frac{[Cl^-]_o}{[Cl^-]_i}\right)^2\right), \tag{20}$$

where Z_{Cl} is the valence of Cl^-, I_{NaKCl} is a cotransport current coefficient, and $[Na^+]$, $[K^+]$ and $[Cl^-]$ are the corresponding concentrations outside and inside (subscripts o and i, respectively) of the cell. $R_{NaKCl,cGMP}$ represents the cGMP-dependent regulation factor that is defined as:

$$R_{NaKCl,cGMP} = 1 + f_{NaKCl,cGMP} \frac{[cGMP]}{[cGMP] + K_{cGMP,NaKCl}}, \tag{21}$$

where $f_{NaKCl,cGMP}$ is a fold-increase in cotransport current due to cGMP. All parameters and their values are presented in **Table 5**.

Very little is known about the effect of cGMP on the NKCC. The model is more or less built on one single reference [49], which also offers limited information for

Parameter	Description	Values [51]
I_{NaKCl}	Cotransport current coefficient	0.106 pA
$f_{NaKCl,cGMP}$	Additional fold-increase in electric current due to cGMP	3.5
$K_{cGMP,NaKCl}$	Half-saturation constant in cGMP-dependent factor	6.4 μM

Table 5.
Parameter values for the cGMP-dependent $Na^+/K^+/Cl^-$ cotransport (NKCC) current.

determining the reliable parameter values. The knowledge of the overall impact of NKCC on VSMC contraction is lacking. Hence, their inclusion in the cGMP-dependent mechanisms seems speculative.

4.6 The model of cGMP-mediated Ca^{2+} flux through the sarco−/endo-plasmic reticulum Ca^{2+}-ATPase (SERCA)

Here we present a novel model of cGMP-dependent activation of the SERCA pump based on the solid-state NMR spectroscopy measurements [35] and the measurements performed on the isolated gastric SMC [36]. The former experiment [35] revealed the physical interactions between the SERCA and the PLB in either a phosphorylated or dephosphorylated state, and the latter experiment [36] offered the results on the increase in Ca2+ uptake as a function of cGMP. The experiments performed on isolated lipid bilayer-bound proteins revealed that the PLB-dependent SERCA activity regulation is allosteric and that SERCA activity depends on the transient conformational equilibrium states of PLB [35]. It was found that phosphorylation at Ser16 of PLB shifts the conformation of PLB towards a more extended and SERCA-bound state, which is non-inhibitory [35]. Phosphorylation of PLB was induced by β-adrenergic stimulation, and it was supposed that the phosphorylation was cAMP/PKA dependent [35]. However, the cGMP/PKG-I dependent phosphorylation of PLB at Ser16 in contact with SERCA was previously shown *in vitro* [33]. Gustavsson et al. [35] proposed that PLB does not function as a simple on/off switch of SERCA. Still, its different conformational equilibrium states exert a gradual control on SERCA activity. PLB phosphorylation does not cause complete dissociation of PLB from SERCA, but it influences the conformational equilibrium of PLB's regulatory domain and shifts its populations towards the non-inhibitory state. That relieves the inhibition of SERCA [35]. According to [35], different PLB/SERCA states exhibit functioning that follows Michaelis–Menten kinetics with the same Hill coefficient (n) and same (V_{max}) but different half-saturation constant (K_m), which is lower for higher relaxation-agonist level. Since the effect of cGMP solely on the half-saturation constant could not explain the increase in Ca^{2+} uptake as a function of high cGMP concentration that was observed *in vitro* in gastric SMC [36], we upgraded the model also by adding a cGMP-dependent regulatory factor into the parameter V_{max} of the standard Michaelis–Menten kinetics, which for SERCA reads:

$$J_{SERCA} = V_{SERCA,min} R_{SERCA,cGMP} \frac{\left[Ca^{2+}\right]_i^{n_{SERCA,Ca}}}{\left[Ca^{2+}\right]_i^{n_{SERCA,Ca}} + R_{Ca,cGMP}^{n_{SERCA,Ca}} K_{Ca,SERCA,max}^{n_{SERCA,Ca}}}, \tag{22}$$

where $R_{SERCA,cGMP}$ is a cGMP-dependent pumping rate regulatory factor, which is according to [36] an increasing Hill function superimposed on the basal pumping rate $V_{SERCA,min}$. The Hill function represents the best fit to the measured data of cGMP dependent increase in Ca^{2+} uptake [36]:

$$R_{SERCA,cGMP} = 1 + f_{SERCA,cGMP} \frac{[cGMP]^{n_{SERCA,cGMP,V}}}{[cGMP]^{n_{SERCA,cGMP,V}} + K_{cGMP,SERCA,V}^{n_{SERCA,cGMP,V}}}. \tag{23}$$

$R_{Ca,cGMP}$ is a cGMP-dependent half-saturation constant regulatory factor in the Eq. (22), which is according to the measurements [35] a decaying Hill function:

$$R_{Ca,cGMP} = 1 - f_{Ca,cGMP} \frac{[cGMP]^{n_{SERCA,cGMP,K}}}{[cGMP]^{n_{SERCA,cGMP,K}} + K_{cGMP,SERCA,K}^{n_{SERCA,cGMP,K}}}. \tag{24}$$

Parameter	Description	Value	References
$n_{SERCA,Ca}$	Hill coefficient	2.5	[50]
$V_{SERCA,min}$	Minimal Ca^{2+} pumping rate	1.88×10^3 μM/s	[50]
$f_{SERCA,cGMP}$	Additional fold increase in SERCA activity due to cGMP	1.44	Recalculated by fitting from [36]
$K_{cGMP,SERCA,V}$	Half-saturation constant in the cGMP-dependent regulatory Hill function	1.44×10^2 μM	Recalculated by fitting from [36]
$n_{SERCA,cGMP,V}$	Hill coefficient	0.092	Recalculated by fitting from [36]
$K_{Ca,SERCA,max}$	Maximal value of SERCA half-saturation constant	0.07 μM	[50]
$f_{Ca,cGMP}$	Additional fold decrease in SERCA activity due to cGMP	70	Recalculated by fitting from [36]
$K_{cGMP,SERCA,K}$	Half saturation constant in the cGMP-dependent regulatory Hill function	0.1 μM	Recalculated by fitting from [35]
$n_{SERCA,cGMP,K}$	Hill coefficient	1.2	Recalculated by fitting from [35]

Table 6.
Parameter values for the cGMP-dependent Ca^{2+} efflux via SERCA.

All parameter values and their descriptions are presented in **Table 6**.

It has to be noted that the parameter values for Eq. (25) were determined by the best fit to only three measured values from [35], and that $K_{SERCA,Ca,max}$ is considered the same as in the existing model for VSMC [50]. In the case of $V_{SERCA,min}$, the best fit was done to five measured points [36].

The significance of the cGMP effect on SERCA is still debated, and it is challenging to consider it independently of other $[Ca^{2+}]_i$-off mechanisms. It is suggested [78] that cGMP-dependent SERCA activity can play a significant role in modulating smooth muscle $[Ca^{2+}]_i$, but its role in the cGMP-mediated relaxation is minor. Therefore, it would be worth testing the significance of that mechanism on the whole-cell-like VSMC model.

4.7 The model of cGMP-mediated current through the plasma membrane Ca^{2+}-ATPase (PMCA)

Yoshida et al. [40] demonstrated that PKG phosphorylated and stimulated PMCA in a concentration-dependent manner. The experiment was conducted on isolated and purified PMCA from the porcine aorta. Much smaller - physiological cGMP concentration, 1 μM, than in previous experiments (500 μM) [38, 39], was added to 10 μg/mL (roughly 0.2 μM) PKG at different free Ca^{2+} concentrations. That increased PMCA activity by approximately 3-fold over the whole range of Ca^{2+} concentrations and slightly shifted the pumping activity towards the left. cGMP alone did not affect the pump activity [40]. In modeling these effects, we use a similar approach as for SERCA, which obeys Michaelis–Menten kinetics. However, previous studies [50, 79] also included weak membrane-potential-dependence, which we also consider here:

$$I_{PMCA} = I_{PMCA,min} R_{PMCA,cGMP} \frac{\left[Ca^{2+}\right]_i^{n_{PMCA,Ca}}}{\left[Ca^{2+}\right]_i^{n_{PMCA,Ca}} + K_{Ca,PMCA}^{n_{PMCA,Ca}}} \left(1 + \frac{V_m - k_\beta}{k_\alpha}\right), \quad (25)$$

Parameter	Description	Value	References
$n_{PMCA,Ca}$	Hill coefficient	0.6	Recalculated by fitting from [40]
$I_{PMCA,min}$	Minimal Ca^{2+} pumping current	0.90 pA	[50]
$K_{Ca,PMCA}$	Half-saturation constant in Ca^{2+}-dependent factor	0.18 μM	Recalculated by fitting from [40]
k_{β}	PMCA voltage sensitivity constant	−100 mV	[50]
k_{α}	PMCA voltage sensitivity constant	250 mV	[50]
$f_{PMCA,cGMP}$	Additional fold-increase in electric current due to cGMP	3	Recalculated by fitting from [40]
$K_{cGMP,PMCA}$	Half-saturation constant in the cGMP-dependent regulatory Hill function	0.50 μM	Recalculated by fitting from [40]
$n_{PMCA,cGMP}$	Hill coefficient	1.7	Recalculated by fitting from [40]

Table 7.
Parameter values for the cGMP-dependent Ca^{2+} current via PMCA.

where $I_{PMCA,min}$ is a minimal pumping rate translated into electric current, which is, according to [40], a function of PKG. Yoshida et al. [40] conducted all their experiments at variable PKG concentrations and 5 to 20 times higher cGMP concentration. Since all our functions were written as cGMP-dependent, we translate PKG concentrations into cGMP by considering the active PKG:cGMP molar ratio 1:4. The Hill function fitted to measured data [40] is superimposed on the basal pumping current $I_{PMCA,min}$ and is here represented as a cGMP-dependent regulatory factor:

$$R_{PMCA,cGMP} = 1 + f_{PMCA,cGMP} \frac{[cGMP]^{n_{PMCA,cGMP}}}{[cGMP]^{n_{PMCA,cGMP}} + K_{cGMP,PMCA}^{n_{PMCA,cGMP}}}. \quad (26)$$

Parameter values and their descriptions are presented in **Table 7**.

In Eq. (25), $K_{Ca,PMCA}$ is a half-saturation constant. The value was determined by fitting the Hill function to two sets of measured data [40], the control case, and the PKG-dependent case, with 0.2 μM PKG and 1 μM cGMP. For the former case, the value is 0.22 μM, and for the latter case, it is 0.14 μM. Since the change is rather small and there are only two measured values available, we do not assume cGMP dependency in this case. That decision is also supported by the results of [39] where the left-shift in that value was only by 27% at supramaximal membrane-permeable cGMP analog concentration (500 μM). We propose here the average value. Elsewhere the value is similar (0.2 μM) [50, 79] and 0.17 μM [51]. $n_{PMCA,Ca}$ is also determined by fitting to the same set of measured data [40]. The values were 0.7 and 0.5 for the control and the PKG-dependent case, respectively. We propose here an average. Value 1 was used elsewhere [50, 51, 79]. Fold-increase in electric current due to cGMP is quite large and might impact the $[Ca^{2+}]_i$, which the whole-cell-like model could demonstrate.

4.8 The model of cGMP-mediated Ca^{2+} flux through the inositol 1,4,5-trisphosphate (IP_3) receptor channels type 1 (IP_3R1)

4.8.1 Variant A

The proposed model for cGMP-mediated IP_3R1 deactivation is also presented here for the first time. The framework of the proposed mechanism is the model of the Ca^{2+} efflux via IP_3R1 as proposed by [50]. That model is upgraded here according to the experimental data of [36], with an additional regulatory factor $R_{IR1,cGMP}$, which accounts for the drop in Ca^{2+} release after IP_3 stimulation with increasing cGMP levels [36]. General description of the Ca^{2+} flux across the SR membrane through the IP_3R1 channels (J_{IR1}) follows [50]:

$$J_{IR1} = g_{IR1} P_{IR1}\left(\left[Ca^{2+}\right]_{SR} - \left[Ca^{2+}\right]_i\right), \tag{27}$$

where g_{IR1} is an overall maximal rate of the channel permeability and P_{IR1} is the channel open probability, which is a biphasic bell-shaped function of $[Ca^{2+}]_i$, and is also dependent on the cytosolic IP_3 and sarcoplasmic Ca^{2+} concentrations ($[IP_3]$ and $[Ca^{2+}]_{SR}$, respectively). It is modeled as in [50, 80]:

$$P_{IR1} = f_{IR1,A}\left(1 - f_{IR1,I}\right) R_{IR1,cGMP} \frac{[IP_3]^{n_{IP3}}}{[IP_3]^{n_{IP3}} + K_{IP3}^{n_{IP3}}} \frac{\left[Ca^{2+}\right]_{SR}^{n_{SR}}}{\left[Ca^{2+}\right]_{SR}^{n_{SR}} + K_{Ca,SR}^{n_{SR}}}. \tag{28}$$

The regulatory factor $R_{IR1,cGMP}$ is a decaying Hill function that depends on the cGMP concentration and is an upgrade from the previous model [50]. The proposed function is the best fit to the measured decay of IP_3-induced Ca^{2+} release as a function of cGMP concentration in isolated gastric SMC [36]:

$$R_{IR1,cGMP} = 1 - f_{IR1,cGMP} \frac{[cGMP]^{n_{IR1,cGMP}}}{[cGMP]^{n_{IR1,cGMP}} + K_{cGMP,IR1}^{n_{IR1,cGMP}}}, \tag{29}$$

where $f_{IR1,cGMP}$ is an additional fold decrease in the channel open probability due to cGMP. $f_{IR1,A}$, and $f_{IR1,I}$ in Eq. (29) are the fractions of the channel population occupied by $[Ca^{2+}]_i$ at the activation sites and inactivation sites, respectively. In this way, the bell-shaped dependency on $[Ca^{2+}]_i$ is achieved. Since activation is a fast process, it is considered to be in the equilibrium:

$$f_{IR1,A} = \frac{\left[Ca^{2+}\right]_i^{n_A}}{\left[Ca^{2+}\right]_i^{n_A} + K_{Ca,A}^{n_A}}, \tag{30}$$

whereas the Ca^{2+}-dependent IP_3R1 inhibition is considered as slow and is therefore modeled with the first-order kinetics as in Eq. (3):

$$\frac{df_{IR1,I}}{dt} = \frac{\bar{f}_{IR1,I} - f_{IR1,I}}{\tau_{IR1,I}}, \tag{31}$$

where $\bar{f}_{IP31,I}$ is the fraction of the inhibited state in equilibrium, which follows a Hill function:

$$\bar{f}_{IR1,I} = \frac{\left[Ca^{2+}\right]_i^{n_I}}{\left[Ca^{2+}\right]_i^{n_I} + K_{Ca,I}^{n_I}}. \tag{32}$$

Description of all parameters and their values are presented in **Table 8**.

4.8.2 Variant B

Other results of Murthy and Zhou [20] provide another possible model description of cGMP-dependent IP_3R1 inhibition. The experiment offers direct PKG or

Parameter	Description	Value	References
g_{IR1}	Maximal permeability rate of the channel	30 s^{-1}	[50]
K_{IP3}	Half saturation constant in the IP_3-dependent regulatory Hill function	0.65 μM	[50]
$K_{Ca,SR}$	Half saturation constant in the sarcoplasmic Ca^{2+}-dependent regulatory Hill function	2×10^3 μM	[50]
n_{IP3}	Hill coefficient	4	[50]
n_{SR}	Hill coefficient	2	[50]
$f_{IR1,cGMP}$	Additional fold decrease in the channel open probability due to cGMP	0.645	Recalculated by fitting from [36]
$K_{cGMP,IR1}$	Half saturation constant in the cGMP-dependent regulatory Hill function	24.6 μM	Recalculated by fitting from [36]
$n_{IR1,cGMP}$	Hill coefficient	0.47	Recalculated by fitting from [36]
$K_{Ca,A}$	Half saturation constant in the cytosolic Ca^{2+}-dependent regulatory Hill function	0.13 μM	[50]
n_A	Hill coefficient	4	[50]
$\tau_{IR1,I}$	Characteristic transition time	6.0 s	[50]
$K_{Ca,I}$	Half saturation constant in the cytosolic Ca^{2+}-dependent regulatory Hill function	0.35 μM	[50]
n_I	Hill coefficient	4	[50]

Table 8.
Parameter values for the cGMP and IP_3 -dependent open probability of IP_3R1 – Variant a.

cGMP dependency of IP_3-dependent Ca^{2+} flux. The cGMP/PKG mediated phosphorylation of IP_3R1 in microsomes was confirmed in the accompanying experiment by immunoprecipitation. Hence, Murthy and Zhou [20] measured Ca^{2+} release through the phosphorylated IP_3R1 within smooth muscle microsomes at different IP_3 concentrations. Prior to measurements, microsomes were either treated with 0.5 μM PKG-Iα holoenzyme and 10 μM cGMP or left intact (control). Ca^{2+} release was determined from the decrease in the steady-state microsomal radioactive Ca^{2+} isotope content. In this way, two dose–response curves were obtained [20]. Their best fits with a Hill function reveal almost the same Hill coefficients (0.49 and 0.42, for the control and cGMP/PKG treated case, respectively) and the same V_{max} (100%) but significantly different half-saturation constants Km, 1.17×10^{-3} μM and 2.35 μM, for the control and the cGMP/PKG treated case, respectively. These two measured values represent two points to which any function could virtually be fitted. Since this is highly unrealistic, we propose the use of competitive, reversible enzyme inhibition kinetics, where cGMP represents an inhibitor in the IP_3-dependent open probability function:

$$OP_{IP3,cGMP} = \frac{[IP_3]^{n_{IP3}}}{[IP_3]^{n_{IP3}} + K_{IP3,cGMP}^{n_{IP3}}}, \tag{33}$$

where $K_{IP3,cGMP}$ is:

$$K_{IP3,cGMP} = K_{IP3,0}\left(1 + \frac{[cGMP]}{K_{i,cGMP}}\right). \tag{34}$$

Parameter	Description	Value	References
$K_{IP3,0}$	Half-saturation constant in the IP_3-dependent regulatory Hill function	0.65 μM	[50]
n_{IP3}	Hill coefficient	4	[50]
$K_{i,cGMP}$	Apparent inhibition constant for cGMP-dependent inhibition of IP_3-dependent open probability of IP_3R1	5.0×10^{-3} μM	Recalculated from [20]
$K_{i,PKG}$	Apparent inhibition constant for PKG-dependent inhibition of IP_3-dependent open probability of IP_3R1	0.25×10^{-3} μM	Recalculated from [20]

Table 9.
Parameter values for the cGMP and IP_3 -dependent open probability of IP_3R1 – Variant B.

The parameter$K_{i,cGMP}$ is recalculated from [20] according to:

$$K_{i,cGMP} = \frac{K_{IP3,c}[cGMP]_0}{K_{IP3,cGMP,i} - K_{IP3,c}}, \tag{35}$$

where $K_{IP3,cGMP,i}$ = 2.35 μM, which is a measured half-saturation constant treated with $[cGMP]_0$ = 10 μM, and $K_{IP3,c}$ = 1.17×10^{-3} μM, which is the corresponding value at control experiment without added cGMP. Eq. (37) gives $K_{i,cGMP}$ = 5.0×10^{-3} μM. The same calculation in which PKG is replacing cGMP in Eqs. (32)–(34) with the value $[PKG]_0$ =0.50 μM yields $K_{i,PKG}$ = 0.25×10^{-3} μM. The summary of parameter values accounting for $OP_{IP3,cGMP/PKG}$ as a function of either cGMP or PKG is presented in **Table 9**.

We offer here two different variants of the mathematical descriptions for the cGMP impact on the IP_3R1. Variant A seems more realistic as it contains the description with saturating Hill function. On the other hand, variant B takes into account the linear relationship on cGMP concentration, which might be questionable at high cGMP concentrations. However, variant B offers an insight into the strength of the inhibition on IP_3R1 exerted by cGMP. $K_{i,cGMP}$ and $K_{i,PKG}$ values indicate very strong inhibition. It is also worth mentioning that such an effect might also arise from the experimental conditions since they were performed on the isolated microsomes [20]. Before actual inclusion of either of both mechanisms into a whole-cell-like model of VSMC, their careful model evaluation at different dynamical levels of $[Ca^{2+}]_i$ signaling, such as membrane potential, basal $[Ca^{2+}]_i$, different shapes and frequencies of $[Ca^{2+}]_i$ signal, would be required.

4.9 The model of cGMP-mediated Ca^{2+}-desensitization of the contractile apparatus

Modeling of cGMP/PKG- dependent Ca^{2+}-desensitization was first introduced by Yang et al. [67], who considered that MLCP is directly activated by cGMP. They modified the 4-state latch bridge model introduced by Hai and Murphy [13] by considering a simple theoretical description of Ca^{2+}/CaM-dependent MLCK activation and MLCP dependent dephosphorylation [67]. They also reduced the model from 4 to 2 states of myosin species, phosphorylated and dephosphorylated (M_p and M, respectively), being in equilibrium. Hence, the relative level of phosphorylated myosin (M_p) was expressed as:

$$M_p = \frac{k_{cat,MLCK}}{k_{cat,MLCK} + R_{MLCP,cGMP} k_{cat,MLCP,b}}, \tag{36}$$

where $k_{cat,MLCK}$ is $[Ca^{2+}]_i$ dependent rate of phosphorylation (see [67]) and, $k_{cat,MLCP,b}$ is a basal dephosphorylation rate that is multiplied by the cGMP-dependent regulatory factor:

$$R_{MLCP,cGMP} = 1 + f_{MLCP,cGMP} \frac{[cGMP]^{n_{MLCP}}}{[cGMP]^{n_{MLCP}} + K_{cGMP,MLCP}^{n_{MLCP}}}, \tag{37}$$

where $f_{cat,MLCP}$ is an additional fold-increase in the MLCP-mediated M_p dephosphorylation rate due to cGMP. $K_{cGMP,MLCP}$ is a half-saturation constant within cGMP-dependent regulatory Hill function with a Hill coefficient n_{MLCP}.

The model of Yang et al. [67] demonstrated cGMP-mediated Ca^{2+}-desensitization by shifting the equilibrium MLC phosphorylation and force curves vs. $[Ca^{2+}]_i$ to the right. However, the model was not used to simulate the time-dependent phosphorylation and force development. In this context, the model would not accurately predict the results since Ca^{2+}-dependent MLCK activation could not be considered as a fast process [81, 82]. Also, the simplification from 4 to 2 states is neither reasonable nor relevant if the model would account for the time-dependent variables. Hence, we propose another modeling approach to tackle the cGMP-dependent activation of MLCP. The proposed model considers the Michaelis–Menten-type of enzyme kinetics for the rate of MLC dephosphorylation within the 4-state latch bridge kinetic scheme [83], yielding the velocity of MLCP dependent dephosphorylation (V_{MLCP}) of both phosphorylated myosin species, attached and detached to actin, AM_p, and M_p, respectively:

$$V_{MLCP} = \frac{d[M_p]}{dt} + \frac{d[AM_p]}{dt} = \frac{R_{MLCP,cGMP} k_{cat,MLCP,b} [MLCP]_{tot}}{[M_p] + [AM_p] + K_{MLCP}} ([M_p] + [AM_p]), \tag{38}$$

where $[MLCP]_{tot}$ is the total MLCP concentration, K_{MLCP} is a Michaelis–Menten constant, and $k_{cat,MLCP}$ is a catalytic rate constant, for which we consider to be cGMP-dependent as proposed by [67] in Eq. (35). All current parameters are presented in **Table 10**.

A similar modeling approach for ROCK-dependent sensitization of the contractile apparatus was used in our previous work [64]. The whole model for all Ca^{2+}/CaM/MLCK interactions, all myosin species, and the time-dependent force development is presented in different variants elsewhere [64, 81, 82, 84] and it comprises more than 12 differential equations. That is a minor drawback of the model, but the model proved itself in describing time-dependent force generation in rat airway smooth muscle cells [64, 82]. Such an extended model would also allow the

Parameter	Description	Value	References
$k_{MLCP,b}$	Basal dephosphorylation rate	$8\ s^{-1}$	[64]
$f_{MLCP,cGMP}$	Additional fold increase in dephosphorylation rate due to cGMP	1	[64]
n_{MLCP}	Hill ceofficient	2	[67]
$K_{cGMP,MLCP}$	Half saturation constant in a cGMP-dependent regulatory Hill function	5.5 μM	[67]
K_{MLCP}	Michaelis–Menten constant	10 μM	[81]
$[MLCP]_{tot}$	Total MLCP concentration	2 μM	[81]

Table 10.
Parameter values for cGMP-dependent Ca^{2+}-desensitization of the contractile apparatus.

modeling of other cGMP/PKG-mediated mechanisms of MLCP and MLCK regulation by considering several different microscopic states of these two enzymes, such as different phosphorylated states, interaction with telokin, CPI-17, etc. That would allow the interconnection of different signal pathways and, hence, the simulation of the effects of various agonists and inhibitors.

5. Conclusion

This work discusses previous and provides the novel cGMP/PKG-dependent mechanisms at the molecular level accounting for their potential use in comprehensive whole-cell-like models of vascular smooth muscle contraction. Much has been done in the fields of measurements and modeling of the cGMP/PKG effects on the individual $[Ca^{2+}]_i$ encoding and decoding mechanisms implicated in VSMC contractility. However, especially in the modeling part, there is still room for improvement and upgrading the existing models and building even multi-cellular [85, 86] and systems-pharmacology based models [87]. We should also take into consideration the importance of coupling the models of vascular smooth muscle cells to endothelial cells that, in response to the shear stress of blood flow, produce NO and other contractile and relaxation mediators [88, 89]. Moreover, the models would enable simulations at the tissue and organ level [90]. However, many of such multi-scale models are weak in describing mechanisms at the molecular level. That is not an easy task since the number of variables and parameters and the model complexity can increase tremendously. The other possibility to tackle that web of interrelated interactions is by complex network approach [91]. However, a dynamic modeling approach, as presented here, which is currently presented only at the level of individual fluxes that need to be assembled into a comprehensive model, offers many more options for studying the temporal dynamical behavior of the system functioning, either under physiological or pathological conditions or after pharmacological intervention. The remarkable advantage and added value of such mathematical models is that they describe the processes as dynamic ones. They often do not consider only one single process but take into account mutual interactions between several highly interrelated variables. In this way, they reach beyond the intuitive thinking of direct and inverse proportions between certain variables, which is often the case when interpreting the experimental results. However, models hide other pitfalls, such as excessive simplicity or complexity, unfounded predictions, prejudging, unawareness of the model's limitations, and transfer of models between different cell types and organisms, and much more. Nevertheless, they represent a useful tool for in-depth insight into the system's dynamical functioning, distinguishing essential from nonessential mechanisms, and last but not least, for highlighting the targets of pharmacological intervention.

Acknowledgements

The author acknowledges the support of the Slovenian Research Agency (ARRS) grant P1–0055.

Conflict of interest

The author declares no conflict of interest.

Author details

Aleš Fajmut
Faculty of Natural Sciences and Mathematics and Faculty of Health Sciences, University of Maribor, Maribor, Slovenia

*Address all correspondence to: ales.fajmut@um.si

References

[1] Kots AY, Martin E, Sharina IG, Murad F. A short history of cGMP, guanylyl cyclases, and cGMP-dependent protein kinases. In: Schmidt HH, Hofmann F, Stasch J-P, editors. cGMP: Generators, Effectors and Therapeutic Implications. Handbook of Experimental Pharmacology. 191. Heidelberg: Springer; 2009. p. 1-14. DOI:10.1007/978-3-540-68964-5_1.

[2] Golshiri K, Ataei Ataabadi E, Portilla Fernandez EC, Jan Danser A, Roks AJ. The importance of the nitric oxide-cGMP pathway in age-related cardiovascular disease: Focus on phosphodiesterase-1 and soluble guanylate cyclase. Basic & clinical pharmacology & toxicology. 2020;127 (2):67-80. DOI:10.1111/bcpt.13319.

[3] Hofmann F, Feil R, Kleppisch T, Schlossmann J. Function of cGMP-dependent protein kinases as revealed by gene deletion. Physiological reviews. 2006;86(1):1-23. DOI:10.1152/physrev.00015.2005.

[4] Tsai EJ, Kass DA. Cyclic GMP signaling in cardiovascular pathophysiology and therapeutics. Pharmacology & therapeutics. 2009;122 (3):216-238. DOI:10.1016/j.pharmthera.2009.02.009.

[5] Murad F. Cellular signaling with nitric oxide and cyclic GMP. Brazilian journal of medical and biological research. 1999;32(11):1317-1327. DOI: 10.1590/s0100-879x1999001100001.

[6] Ando J, Yamamoto K. Flow detection and calcium signalling in vascular endothelial cells. Cardiovascular research. 2013;99(2):260-268. DOI: 10.1093/cvr/cvt084.

[7] Putney JW, Broad LM, Braun F-J, Lievremont J-P, Bird GSJ. Mechanisms of capacitative calcium entry. Journal of cell science. 2001;114(12):2223-2229.

[8] Martinac B. Mechanosensitive ion channels: molecules of mechanotransduction. Journal of Cell Science. 2004;117(12):2449-2460. DOI: 10.1242/jcs.01232.

[9] Ando J, Yamamoto K. Vascular mechanobiology endothelial cell responses to fluid shear stress. Circulation Journal. 2009;73(11): 1983-1992. DOI:10.1253/circj.cj-09-0583.

[10] Francis SH, Blount MA, Corbin JD. Mammalian cyclic nucleotide phosphodiesterases: molecular mechanisms and physiological functions. Physiological reviews. 2011; 91(2):651-690. DOI:10.1152/physrev.00030.2010.

[11] Hill MA, Meininger GA, Davis MJ, Laher I. Therapeutic potential of pharmacologically targeting arteriolar myogenic tone. Trends in pharmacological sciences. 2009;30(7): 363-374. DOI:10.1016/j.tips.2009.04.008.

[12] Somlyo AP, Somlyo AV. Ca2+ sensitivity of smooth muscle and nonmuscle myosin II: modulated by G proteins, kinases, and myosin phosphatase. Physiological reviews. 2003.

[13] Hai C-M, Murphy RA. Cross-bridge phosphorylation and regulation of latch state in smooth muscle. American Journal of Physiology-Cell Physiology. 1988;254(1):C99-C106. DOI:10.1152/ajpcell.1988.254.1.c99.

[14] Brozovich F, Nicholson C, Degen C, Gao YZ, Aggarwal M, Morgan KG. Mechanisms of vascular smooth muscle contraction and the basis for pharmacologic treatment of smooth muscle disorders. Pharmacological reviews. 2016;68(2):476-532. DOI: 10.1124/pr.115.010652.

[15] Marhl M, Noble D, Roux E. Modeling of molecular and cellular mechanisms involved in Ca2+ signal encoding in airway myocytes. Cell Biochemistry and Biophysics. 2006 2006/11/01;46(3):285-302. DOI: 10.1385/cbb:46:3:285.

[16] Gao Y. Biology of vascular smooth muscle: vasoconstriction and dilatation: Springer; 2017. DOI:10.1007/978-981-10-4810-4.

[17] Kim JJ, Lorenz R, Arold ST, Reger AS, Sankaran B, Casteel DE, et al. Crystal structure of PKG I: cGMP complex reveals a cGMP-mediated dimeric interface that facilitates cGMP-induced activation. Structure. 2016;24 (5):710-720. DOI:10.1016/j.str.2016.03.009.

[18] Bonnevier J, Fässler R, Somlyo AP, Somlyo AV, Arner A. Modulation of Ca2 + sensitivity by cyclic nucleotides in smooth muscle from protein kinase G-deficient mice. Journal of Biological Chemistry. 2004;279(7):5146-5151. DOI: 10.1074/jbc.m306532200.

[19] Schlossmann J, Ammendola A, Ashman K, Zong X, Huber A, Neubauer G, et al. Regulation of intracellular calcium by a signalling complex of IRAG, IP 3 receptor and cGMP kinase Iβ. Nature. 2000;404 (6774):197-201. DOI:10.1038/35004606.

[20] Murthy KS, Zhou H. Selective phosphorylation of the IP3R-I in vivo by cGMP-dependent protein kinase in smooth muscle. American Journal of Physiology-Gastrointestinal and Liver Physiology. 2003;284(2):G221-G230. DOI:10.1152/ajpgi.00401.2002.

[21] Geiselhöringer A, Werner M, Sigl K, Smital P, Wörner R, Acheo L, et al. IRAG is essential for relaxation of receptor-triggered smooth muscle contraction by cGMP kinase. The EMBO journal. 2004;23(21):4222-4231. DOI: 10.1038/sj.emboj.7600440.

[22] Ammendola A, Geiselhöringer A, Hofmann F, Schlossmann J. Molecular determinants of the interaction between the inositol 1, 4, 5-trisphosphate receptor-associated cGMP kinase substrate (IRAG) and cGMP kinase Iβ. Journal of Biological Chemistry. 2001; 276(26):24153-24159. DOI:10.1074/jbc.m101530200.

[23] Masuda W, Betzenhauser MJ, Yule DI. InsP3R-associated cGMP kinase substrate determines inositol 1, 4, 5-trisphosphate receptor susceptibility to phosphoregulation by cyclic nucleotide-dependent kinases. Journal of Biological Chemistry. 2010;285(48):37927-37938. DOI:10.1074/jbc.m110.168989.

[24] Desch M, Sigl K, Hieke B, Salb K, Kees F, Bernhard D, et al. IRAG determines nitric oxide-and atrial natriuretic peptide-mediated smooth muscle relaxation. Cardiovascular research. 2010;86(3):496-505. DOI: 10.1093/cvr/cvq008.

[25] Tang M, Wang G, Lu P, Karas RH, Aronovitz M, Heximer SP, et al. Regulator of G-protein signaling-2 mediates vascular smooth muscle relaxation and blood pressure. Nature medicine. 2003;9(12):1506-1512. DOI: 10.1038/nm958.

[26] Xia C, Bao Z, Yue C, Sanborn BM, Liu M. Phosphorylation and regulation of G-protein-activated phospholipase C-β3 by cGMP-dependent protein kinases. Journal of Biological Chemistry. 2001; 276(23):19770-19777. DOI:10.1074/jbc.m006266200.

[27] Sun X, Kaltenbronn KM, Steinberg TH, Blumer KJ. RGS2 is a mediator of nitric oxide action on blood pressure and vasoconstrictor signaling. Molecular pharmacology. 2005;67(3): 631-639. DOI:10.1124/mol.104.007724.

[28] Ruiz-Velasco V, Zhong J, Hume JR, Keef KD. Modulation of Ca2+ channels by cyclic nucleotide cross activation of

opposing protein kinases in rabbit portal vein. Circulation Research. 1998;82(5): 557-565. DOI:10.1161/01.res.82.5.557.

[29] Harraz OF, Welsh DG. Protein kinase A regulation of T-type Ca2+ channels in rat cerebral arterial smooth muscle. Journal of cell science. 2013;126 (13):2944-2954. DOI:10.1242/jcs.128363.

[30] Harraz OF, Brett SE, Welsh DG. Nitric oxide suppresses vascular voltage-gated T-type Ca2+ channels through cGMP/PKG signaling. American Journal of Physiology-Heart and Circulatory Physiology. 2014;306 (2):H279-H285. DOI:10.1152/ajpheart.00743.2013.

[31] Chen J, Crossland RF, Noorani MM, Marrelli SP. Inhibition of TRPC1/TRPC3 by PKG contributes to NO-mediated vasorelaxation. American journal of physiology-Heart and circulatory physiology. 2009;297(1):H417-H424. DOI:10.1152/ajpheart.01130.2008.

[32] Tewari K, Simard JM. Sodium nitroprusside and cGMP decrease Ca 2+ channel availability in basilar artery smooth muscle cells. Pflügers Archiv. 1996;433(3):304-311. DOI:10.1007/s004240050281.

[33] Raeymaekers L, Hofmann F, Casteels R. Cyclic GMP-dependent protein kinase phosphorylates phospholamban in isolated sarcoplasmic reticulum from cardiac and smooth muscle. Biochemical Journal. 1988;252 (1):269-273. DOI:10.1042/bj2520269.

[34] Karczewski P, Kelm M, Hartmann M, Schrader J. Role of phospholamban in NO/EDRF-induced relaxation in rat aorta. Life sciences. 1992;51(15):1205-1210. DOI:10.1016/0024-3205(92)90357-u.

[35] Gustavsson M, Verardi R, Mullen DG, Mote KR, Traaseth NJ, Gopinath T, et al. Allosteric regulation of SERCA by phosphorylation-mediated conformational shift of phospholamban. Proceedings of the National Academy of Sciences. 2013;110(43):17338-17343. DOI:10.1073/pnas.1303006110.

[36] Murthy K, Severi C, Grider J, Makhlouf G. Inhibition of IP3 and IP3-dependent Ca2+ mobilization by cyclic nucleotides in isolated gastric muscle cells. American Journal of Physiology-Gastrointestinal and Liver Physiology. 1993;264(5):G967-G974. DOI:10.1152/ajpgi.1993.264.5.g967.

[37] Furukawa Kl, Ohshima N, Tawada-Iwata Y, Shigekawa M. Cyclic GMP stimulates Na+/Ca2+ exchange in vascular smooth muscle cells in primary culture. Journal of Biological Chemistry. 1991;266(19):12337-12341. DOI:10.1016/s0021-9258(18)98901-5.

[38] Vrolix M, Raeymaekers L, Wuytack F, Hofmann F, Casteels R. Cyclic GMP-dependent protein kinase stimulates the plasmalemmal Ca2+ pump of smooth muscle via phosphorylation of phosphatidylinositol. Biochemical Journal. 1988;255(3):855-863. DOI: 10.1042/bj2550855.

[39] Furukawa K, Tawada Y, Shigekawa M. Regulation of the plasma membrane Ca2+ pump by cyclic nucleotides in cultured vascular smooth muscle cells. Journal of Biological Chemistry. 1988;263(17):8058-8065. DOI:10.1016/s0021-9258(18)68441-8.

[40] Yoshida Y, Sun H-T, Cai J-Q, Imai S. Cyclic GMP-dependent protein kinase stimulates the plasma membrane Ca2+ pump ATPase of vascular smooth muscle via phosphorylation of a 240-kDa protein. Journal of Biological Chemistry. 1991;266(29):19819-19825. DOI:10.1016/s0021-9258(18)55065-1.

[41] Manoury B, Idres S, Leblais V, Fischmeister R. Ion channels as effectors of cyclic nucleotide pathways: Functional relevance for arterial tone

regulation. Pharmacology & therapeutics. 2020;209:107499. DOI: 10.1016/j.pharmthera.2020.107499.

[42] Schubert R, Nelson MT. Protein kinases: tuners of the BKCa channel in smooth muscle. Trends in pharmacological sciences. 2001;22(10): 505-512. DOI:10.1016/s0165-6147(00)01775-2.

[43] Alioua A, Tanaka Y, Wallner M, Hofmann F, Ruth P, Meera P, et al. The large conductance, voltage-dependent, and calcium-sensitive K+ channel, Hslo, is a target of cGMP-dependent protein kinase phosphorylation in vivo. Journal of Biological Chemistry. 1998;273(49): 32950-32956. DOI:10.1074/jbc.273.49.32950.

[44] Zhou X-B, Ruth P, Schlossmann J, Hofmann F, Korth M. Protein phosphatase 2A is essential for the activation of Ca2+-activated K+ currents by cGMP-dependent protein kinase in tracheal smooth muscle and Chinese hamster ovary cells. Journal of Biological Chemistry. 1996;271(33):19760-19767. DOI:10.1074/jbc.271.33.19760.

[45] Fellner SK, Arendshorst WJ. Complex interactions of NO/cGMP/PKG systems on Ca2+ signaling in afferent arteriolar vascular smooth muscle. American Journal of Physiology-Heart and Circulatory Physiology. 2010;298(1):H144-H151. DOI:10.1152/ajpheart.00485.2009.

[46] Sausbier M, Arntz C, Bucurenciu I, Zhao H, Zhou X-B, Sausbier U, et al. Elevated blood pressure linked to primary hyperaldosteronism and impaired vasodilation in BK channel–deficient mice. Circulation. 2005;112(1): 60-68. DOI:10.1161/01.cir.0000156448.74296.fe.

[47] Matchkov VV, Aalkjaer C, Nilsson H. A cyclic GMP–dependent calcium-activated chloride current in smooth-muscle cells from rat mesenteric resistance arteries. Journal of General Physiology. 2004;123(2): 121-134. DOI:10.1085/jgp.200308972.

[48] Tamaoki J, Tagaya E, Nishimura K, Isono K, Nagai A. Role of Na+-K+ ATPase in cyclic GMP-mediated relaxation of canine pulmonary artery smooth muscle cells. British journal of pharmacology. 1997;122(1):112-116. DOI:10.1038/sj.bjp.0701351.

[49] O'Donnell ME, Owen N. Role of cyclic GMP in atrial natriuretic factor stimulation of Na+, K+, Cl-cotransport in vascular smooth muscle cells. Journal of Biological Chemistry. 1986;261(33): 15461-15466. DOI:10.1016/s0021-9258(18)66734-1.

[50] Jacobsen JCB, Aalkjær C, Nilsson H, Matchkov VV, Freiberg J, Holstein-Rathlou N-H. Activation of a cGMP-sensitive calcium-dependent chloride channel may cause transition from calcium waves to whole cell oscillations in smooth muscle cells. American Journal of Physiology-Heart and Circulatory Physiology. 2007;293(1):H215-H228. DOI:10.1152/ajpheart.00726.2006.

[51] Kapela A, Bezerianos A, Tsoukias NM. A mathematical model of Ca2+ dynamics in rat mesenteric smooth muscle cell: agonist and NO stimulation. Journal of theoretical biology. 2008;253(2):238-260. DOI: 10.1016/j.jtbi.2008.03.004.

[52] Kiss A, Erdődi F, Lontay B. Myosin phosphatase: Unexpected functions of a long-known enzyme. Biochimica et Biophysica Acta (BBA)-Molecular Cell Research. 2019;1866(1):2-15. DOI: 10.1016/j.bbamcr.2018.07.023.

[53] Kimura K, Ito M, Amano M, Chihara K, Fukata Y, Nakafuku M, et al. Regulation of myosin phosphatase by Rho and Rho-associated kinase (Rho-kinase). Science. 1996;273(5272): 245-248. DOI:10.1126/science.273.5272.245.

[54] Wooldridge AA, MacDonald JA, Erdodi F, Ma C, Borman MA, Hartshorne DJ, et al. Smooth muscle phosphatase is regulated in vivo by exclusion of phosphorylation of threonine 696 of MYPT1 by phosphorylation of Serine 695 in response to cyclic nucleotides. Journal of Biological Chemistry. 2004;279(33): 34496-34504. DOI:10.1074/jbc.m405957200.

[55] Gao Y, Portugal AD, Negash S, Zhou W, Longo LD, Usha Raj J. Role of Rho kinases in PKG-mediated relaxation of pulmonary arteries of fetal lambs exposed to chronic high altitude hypoxia. American Journal of Physiology-Lung Cellular and Molecular Physiology. 2007;292(3):L678-L684. DOI:10.1152/ajplung.00178.2006.

[56] Khromov A, Choudhury N, Stevenson AS, Somlyo AV, Eto M. Phosphorylation-dependent autoinhibition of myosin light chain phosphatase accounts for Ca2+ sensitization force of smooth muscle contraction. Journal of Biological Chemistry. 2009;284(32):21569-21579. DOI:10.1074/jbc.m109.019729.

[57] Terrak M, Kerff F, Langsetmo K, Tao T, Dominguez R. Structural basis of protein phosphatase 1 regulation. Nature. 2004;429(6993):780-784. DOI: 10.1038/nature02582.

[58] Grassie ME, Sutherland C, Ulke-Lemée A, Chappellaz M, Kiss E, Walsh MP, et al. Cross-talk between Rho-associated kinase and cyclic nucleotide-dependent kinase signaling pathways in the regulation of smooth muscle myosin light chain phosphatase. Journal of Biological Chemistry. 2012; 287(43):36356-36369. DOI:10.1074/jbc.m112.398479.

[59] Lubomirov L, Papadopoulos S, Filipova D, Baransi S, Todorović D, Lake P, et al. The involvement of phosphorylation of myosin phosphatase targeting subunit 1 (MYPT 1) and MYPT 1 isoform expression in NO/cGMP mediated differential vasoregulation of cerebral arteries compared to systemic arteries. Acta Physiologica. 2018;224(1):e13079. DOI: 10.1111/apha.13079.

[60] Eto M. Regulation of cellular protein phosphatase-1 (PP1) by phosphorylation of the CPI-17 family, C-kinase-activated PP1 inhibitors. Journal of Biological Chemistry. 2009; 284(51):35273-35277. DOI:10.1074/jbc.r109.059972.

[61] Smolenski A, Lohmann SM, Bertoglio J, Chardin P, Sauzeau V, Le Jeune H, et al. Cyclic GMP-dependent protein kinase signaling pathway inhibits RhoA-induced Ca2+ sensitization of contraction in vascular smooth muscle. Journal of Biological Chemistry. 2000;275(28):21722-21729. DOI:10.1074/jbc.m000753200.

[62] Roux E, Mbikou P, Fajmut A. Role of protein kinase network in excitation-contraction coupling in smooth muscle cell. Protein Kinases. 2012:287-320. DOI: 10.5772/37805.

[63] Butler T, Paul J, Europe-Finner N, Smith R, Chan E-C. Role of serine-threonine phosphoprotein phosphatases in smooth muscle contractility. American Journal of Physiology-Cell Physiology. 2013;304(6):C485-C504. DOI:10.1152/ajpcell.00161.2012.

[64] Mbikou P, Fajmut A, Brumen M, Roux E. Contribution of Rho kinase to the early phase of the calcium–contraction coupling in airway smooth muscle. Experimental physiology. 2011; 96(2):240-258. DOI:10.1113/expphysiol.2010.054635.

[65] Wu X, Haystead TA, Nakamoto RK, Somlyo AV, Somlyo AP. Acceleration of myosin light chain dephosphorylation and relaxation of smooth muscle by telokin: synergism with cyclic

nucleotide-activated kinase. Journal of Biological Chemistry. 1998;273(18): 11362-11369. DOI:10.1074/jbc.273.18.11362.

[66] Khromov AS, Momotani K, Jin L, Artamonov MV, Shannon J, Eto M, et al. Molecular mechanism of telokin-mediated disinhibition of myosin light chain phosphatase and cAMP/cGMP-induced relaxation of gastrointestinal smooth muscle. Journal of Biological Chemistry. 2012;287(25):20975-20985. DOI:10.1074/jbc.m112.341479.

[67] Yang J, Clark JW, Bryan RM, Robertson CS. Mathematical modeling of the nitric oxide/cGMP pathway in the vascular smooth muscle cell. American Journal of Physiology-Heart and Circulatory Physiology. 2005;289(2): H886-H897. DOI:10.1152/ajpheart.00216.2004.

[68] Yang J, Clark Jr JW, Bryan RM, Robertson C. The myogenic response in isolated rat cerebrovascular arteries: smooth muscle cell model. Medical engineering & physics. 2003;25(8): 691-709. DOI:10.1016/s1350-4533(03)00100-0.

[69] Zhou X-B, Arntz C, Kamm S, Motejlek K, Sausbier U, Wang G-X, et al. A molecular switch for specific stimulation of the BKCa channel by cGMP and cAMP kinase. Journal of Biological Chemistry. 2001;276(46): 43239-43245. DOI:10.1074/jbc.m104202200.

[70] Mistry D, Garland C. Nitric oxide (NO)-induced activation of large conductance Ca2+-dependent K+ channels (BKCa) in smooth muscle cells isolated from the rat mesenteric artery. British journal of pharmacology. 1998; 124(6):1131-1140. DOI:10.1038/sj.bjp.0701940.

[71] Lee MR, Li L, Kitazawa T. Cyclic GMP causes Ca2+ desensitization in vascular smooth muscle by activating the myosin light chain phosphatase. Journal of Biological Chemistry. 1997; 272(8):5063-5068. DOI:10.1074/jbc.272.8.5063.

[72] Stockand JD, Sansom SC. Mechanism of activation by cGMP-dependent protein kinase of large Ca (2+)-activated K+ channels in mesangial cells. American Journal of Physiology-Cell Physiology. 1996;271(5):C1669-C1677. DOI:10.1152/ajpcell.1996.271.5.c1669.

[73] Large WA, Wang Q. Characteristics and physiological role of the Ca (2+)-activated Cl-conductance in smooth muscle. American Journal of Physiology-Cell Physiology. 1996;271 (2):C435-C454. DOI:10.1152/ajpcell.1996.271.2.c435.

[74] Jacobsen JCB, Aalkjær C, Nilsson H, Matchkov VV, Freiberg J, Holstein-Rathlou N-H. A model of smooth muscle cell synchronization in the arterial wall. American Journal of Physiology-Heart and Circulatory Physiology. 2007;293 (1):H229-H237. DOI:10.1152/ajpheart.00727.2006.

[75] Di Francesco D, Noble D. A model of cardiac electrical activity incorporating ionic pumps and concentration changes. Philosophical Transactions of the Royal Society of London B, Biological Sciences. 1985;307 (1133):353-398. DOI:10.1098/rstb.1985.0001.

[76] Fotis H, Tatjanenko LV, Vasilets LA. Phosphorylation of the α-subunits of the Na+/K+-ATPase from mammalian kidneys and Xenopus oocytes by cGMP-dependent protein kinase results in stimulation of ATPase activity. European journal of biochemistry. 1999; 260(3):904-910. DOI:10.1046/j.1432-1327.1999.00237.x.

[77] Strieter J, Stephenson JL, Palmer LG, Weinstein AM. Volume-activated chloride permeability can

mediate cell volume regulation in a mathematical model of a tight epithelium. The Journal of general physiology. 1990;96(2):319-344. DOI: 10.1085/jgp.96.2.319.

[78] Lalli MJ, Shimizu S, Sutliff RL, Kranias EG, Paul RJ. [Ca2+] ihomeostasis and cyclic nucleotide relaxation in aorta of phospholamban-deficient mice. American Journal of Physiology-Heart and Circulatory Physiology. 1999;277(3):H963-H970. DOI:10.1152/ajpheart.1999.277.3.h963.

[79] Parthimos D, Edwards DH, Griffith T. Minimal model of arterial chaos generated by coupled intracellular and membrane Ca2+ oscillators. American Journal of Physiology-Heart and Circulatory Physiology. 1999;277 (3):H1119-H1144. DOI:10.1152/ajpheart.1999.277.3.h1119.

[80] Imtiaz MS, Smith DW, van Helden DF. A theoretical model of slow wave regulation using voltage-dependent synthesis of inositol 1, 4, 5-trisphosphate. Biophysical journal. 2002;83(4):1877-1890. DOI:10.1016/s0006-3495(02)73952-0.

[81] Fajmut A, Brumen M. MLC-kinase/phosphatase control of Ca2+ signal transduction in airway smooth muscles. Journal of theoretical biology. 2008;252 (3):474-481. DOI:10.1016/j.jtbi.2007.10.005.

[82] Mbikou P, Fajmut A, Brumen M, Roux E. Theoretical and experimental investigation of calcium-contraction coupling in airway smooth muscle. Cell biochemistry and biophysics. 2006;46 (3):233-251. DOI:10.1385/cbb:46:3:233.

[83] Fajmut A, Dobovišek A, Brumen M. Mathematical modeling of the relation between myosin phosphorylation and stress development in smooth muscles. Journal of chemical information and modeling. 2005;45(6):1610-1615. DOI: 10.1021/ci050178a.

[84] Fajmut A, Brumen M, Schuster S. Theoretical model of the interactions between Ca2+, calmodulin and myosin light chain kinase. FEBS letters. 2005; 579(20):4361-4366. DOI:10.1016/j.febslet.2005.06.076.

[85] Kapela A, Nagaraja S, Tsoukias NM. A mathematical model of vasoreactivity in rat mesenteric arterioles. II. Conducted vasoreactivity. American Journal of Physiology-Heart and Circulatory Physiology. 2010;298(1): H52-H65. DOI:10.1152/ajpheart.00546.2009.

[86] Koenigsberger M, Sauser R, Meister J-J. Emergent properties of electrically coupled smooth muscle cells. Bulletin of mathematical biology. 2005; 67(6):1253-1272. DOI:10.1016/j.bulm.2005.02.001.

[87] Garmaroudi FS, Handy DE, Liu Y-Y, Loscalzo J. Systems Pharmacology and Rational Polypharmacy: Nitric Oxide− Cyclic GMP Signaling Pathway as an Illustrative Example and Derivation of the General Case. PLoS computational biology. 2016;12(3): e1004822. DOI:10.1371/journal.pcbi.1004822.

[88] Koo A, Nordsletten D, Umeton R, Yankama B, Ayyadurai S, García-Cardeña G, et al. In silico modeling of shear-stress-induced nitric oxide production in endothelial cells through systems biology. Biophysical journal. 2013;104(10):2295-2306. DOI:10.1016/j.bpj.2013.03.052.

[89] Sriram K, Laughlin JG, Rangamani P, Tartakovsky DM. Shear-induced nitric oxide production by endothelial cells. Biophysical journal. 2016;111(1):208-221. DOI:10.1016/j.bpj.2016.05.034.

[90] Comerford A, Plank M, David T. Endothelial nitric oxide synthase and calcium production in arterial geometries: an integrated fluid

mechanics/cell model. Journal of biomechanical engineering. 2008;130 (1). DOI:10.1115/1.2838026.

[91] Gosak M, Markovič R, Dolenšek J, Rupnik MS, Marhl M, Stožer A, et al. Network science of biological systems at different scales: A review. Physics of life reviews. 2018;24:118-135. DOI:10.1016/j.plrev.2017.11.003.